INVENTORSHIP

INVENTORSHIP

The Art of Innovation

LEONARD M. GREENE

JOHN WILEY & SONS, INC.

New York • Chichester • Weinheim • Brisbane • Singapore • Toronto

This book is dedicated to Hugo Gernsback, an inventor and science fiction writer, and a lifelong inspiration to me. His compelling (and, as time has proven, uncannily accurate) visions of the future sparked the imagination and spurred the ambitions of a very young Leonard Greene.

If Elmer Sperry showed aircraft pilots the way home, it was Leonard Greene who ensured our safe arrival.

SENATOR JOHN GLENN

Necessity is not the mother of invention; Imagination is.

LEONARD M. GREENE

CONTENTS

FOREWORD

Here is a most fascinating volume written by one of perhaps a handful of people qualified to do so. Leonard Greene is not only a skillful writer, but he is a prime example of the innovator/inventor—a calling that Leonard has compressed innovatively into a new term: Inventorship.

From this lively and personal account, we learn that we can all practice inventorship to great advantage—measured either in gold or pleasure, or perhaps both! Leonard's patented inventions now run well over a hundred and still counting—with several new ones in 2000 alone. There is scarcely a flying machine built today that does not include his devices to enhance safety. And surely thousands of lives have been saved through their use. Gold and pleasure, indeed!

Leonard suggests that we are all capable of inventorship. It begins simply enough, he says, if we use our imaginations. Inventions, he makes us realize, need not all be the product of great scientific or technological breakthroughs. He doesn't say it in so many words, but his innovative idea is that just the act of, say, contemplating an envelope with the goal of eliminating the necessity to lick same, is the first step toward successful inventorship.

His book is full of fascinating examples that illustrate this premise. So well does it accomplish its purpose of awakening in all of us our capacity for inventorship, that it sent me back through a cascade of memories of innovative thinking that served my colleagues and me well in the profession of journalism.

An example: As the Russian troops and those of the western Allies met in central Europe at the close of World War II, we war correspondents were anxious to get into the areas occupied by the Russian troops. The Russian guards, however, were demanding to see credentials authorizing our entry, and the Russian high command wasn't issuing any such passes.

Then one day, one of our number, Jimmy Cannon, came back to our Third Army press camp after two days away, and announced triumphantly that he had been in the Russian sector. He had shown the guards his Texaco credit card with the oil company's big red star on its back. The guards couldn't read English, but that Soviet symbol was enough, and pass him through they did.

That, as Leonard Greene has named it, is inventorship.

Walter Cronkite

ACKNOWLEDGMENTS

I would like to thank the many individuals who contributed to making this book possible:

Fintan O'Hare, who managed to visualize all of my ideas and translate them into his unique illustrations.

At the Institute for SocioEconomic Studies: Thomas M. Cassidy for his expertise on elder care and health care issues, Helen Gibbs for her keen questions and polished editing, Bonnie Le Var for her detailed memory and many reminders, Allan Ostergren for his perceptions and attention to the big picture.

At Safe Flight Instrument Corporation: Mary Blancato for her tireless daily management of my busy schedule, and Peter Fleiss for his businesslike attention to specifics.

My agent, Teresa Hartnett; my book project consultant, Donald O. Graul Jr.; and my editor, Michael J. Hamilton, also deserve special thanks.

This book is in many ways a product of my innovative and inventive life and I would like to thank my wife Joyce and my entire family for helping me to enjoy every minute of it.

L.M.G.

INTRODUCTION
INVENTORSHIP: THE ART
OF INNOVATION

Imagination is more important than knowledge, for knowledge is limited, whereas imagination embraces the entire world.

—ALBERT EINSTEIN

From early childhood, all of us have an innate potential to be inventive. And that is the reason you should read this book: to learn to use what you already know to improve your life, and to encourage your children to have richer, fuller, more interesting lives. My purpose is to demonstrate to you, in the form of anecdotes, the process of what I call *inventorship*. I believe that by reading this book you will recognize the inventor in you, in your children, in all of us. Inventorship is not unique. If I can practice it, so can you.

You will not find information here about how to register your ideas for patents, or how to make a million dollars by discovering the next innovative technology. My goal is to help you understand that once you recognize, then develop, your innate inventorship capability, you will be able to use it to find new and better ways to solve everyday problems and to achieve your professional and personal goals. Inventorship is not exotic or unfathomable; in fact, it can be defined simply as learning to develop the process of finding another way. To be more specific, I should say "the process of finding an *other* way," because the goal is not just to do something over again, but to find an alternative approach to a problem. Let me give you the first of the many examples I use throughout the book to clarify this important point.

Late in the twentieth century, the National Aeronautics and Space Administration (NASA) sent a probe to Mars that crashed into the Red Planet and was destroyed. A brief investigation turned up the reason. One group of engineers was using the metric system (centimeters/grams) of measure and a second group, working on another part of the project, was using the imperial (feet/pounds) system, but neither group told the other. If "an other" method had been used—if one group purposely used metric and the other did the same calculations using imperial, and the results were compared to see if they agreed—the mission would have been successful. Very often, inventorship simply requires looking for and finding another way.

Some may argue with my premise in this book; they may claim that inventors are made, not born. To make my case, I suggest all we need do is watch the actions of our children. Can there be any doubt we are all born with a marvelous capacity to be inventive? I can trace many of my best ideas back to my own childhood experiences. The result of following my inventive nature is that I now hold more than 100 patents, not only in aeronautics, my primary field of study, but also in such diverse areas as skiing, sailing, and chess. And I'm proud to say I am one of the 135 members of the National Inventors Hall of Fame. I tell you that not to impress you, but to assure you that I have the credentials for writing a book on inventorship. I want you to be able to trust that what I say is true. I am not somehow special or different from you; I was just fortunate enough to follow the wanderings of my inventive mind. I'd like to help you do the same.

My career, and the careers of many of my colleagues, provide abundant evidence that we can all improve our lives and those of our fellow human beings in a host of different areas—and have a lot of fun while we're at it—if we just look around us with curiosity and imagination, and then find another way.

INVENTORSHIP

1

The Nature of Inventorship

Chance favors the prepared mind.

—LOUIS PASTEUR (1822–1895)

Ask people about invention, innovation, or creativity, and their answers will reveal several misconceptions:

- The process is mysterious.
- Only a very few gifted people have the talent to be inventors.
- Great discoveries are often the result of luck or accident.
- Inventors must have scientific training.
- To be inventive, you have to work in a "special environment," such as a research laboratory.

In truth, most people are born with a wonderful capability to invent. But before I explain how you can tap into and fully develop this capacity, and then apply it to such efforts as starting a business or raising your children, let's discuss what inventorship is all about.

To begin with, let's rid ourselves of the common notion that an invention necessarily leads to that magical-sounding document called a patent. There's a big difference between inventions and patents. You can be inventive and never take out a patent. That distinction is exciting because it widens the view of the inventorship process to include being innovative. I believe that innovative thoughts come to mind once you have developed the skill of innovation, and an innovative mind naturally tends to be inventive.

To the prepared mind, innovation and imagination are spontaneous acts. An imaginative thought can occur in a fraction of a second. Such a thought may occur as a picture. It's present in your mind, waiting for the split second of your awakening to see it. Awakening to it, often as if from a daydream, will be your insight. Typically, such a thought, or picture, emerges complete. The time element is only the time it takes for you to express the view (or thought) in whatever form you choose. To be prepared to capture your imagination, you need only learn the process of inventorship.

Be Ready to Discover

An apple falls on Sir Isaac Newton's head, and—*voilà!*—he discovers gravity. That's an example of the folklore of invention. It suggests that scientific discoveries are often the result of happenstance or accident, and if it weren't for random events, many of the advances we enjoy today simply wouldn't exist. Nothing could be further from the truth.

Yes, a falling apple inspired Newton to make his revelatory discovery, but inspiration is not synonymous with accident. When the apple fell, Newton was ready to recognize the meaning behind it. His mind had been trained to see important relationships in seemingly ordinary occurrences. (See Figure 1.1.) Apples—and doubtless many other things—had

Figure 1.1 When the apple fell on Sir Isaac Newton's head, his mind was ready to make the connections necessary to formulate his theory of gravity.

been falling on people's heads long before the fateful fruit landed on Sir Isaac. Why didn't any of those other people have the same realization as Newton? Their minds weren't ready; they hadn't been prepared.

Our everyday experiences contain many secrets, but they reveal themselves only to those who have developed the necessary awareness. As an example, suppose two people are daydreaming about fishing. One gets up, finds his fishing pole, buys some bait, and heads for the river to try his luck. The other starts thinking about the process of catching fish and devises a new lure that's more attractive to the fish. He then goes fishing and brings home a record catch. The daydream was the same for both, but one was better prepared to do something inventive with it.

As children, we're taught that daydreaming is a waste of time. Some parents and teachers even label it bad behavior and an indication that we're not paying attention to the important things in life. That's true only if we don't make use of our daydreams by turning them into realities. To minds trained in inventorship, daydreaming is essential. Inventions come about when people are mentally receptive to the thoughts that tell them there is another, often better, way to achieve a result or a goal. Fortunately, being receptive can be learned. So let's get started. If something falls on, or pops into, your head, you'll be prepared to make the most of it.

Defining Inventorship

In my experience, innovations happen in two ways. I call the first *Research Inventorship*. This is what most of us think of when we read or hear about inventions. The image is something like this: A group of scientists in a laboratory spends a great deal of time and money performing a series of experiments. If all goes well, they emerge with an innovation—whether in technology, medicine, aviation, or any one of thousands of other areas of study.

My term for the second way is *Eureka Inventorship*. This is the form of inventorship I focus on in this book. I was inspired to call it that in honor of the Greek mathematician, physicist, and inventor Archimedes (c. 287–212 B.C.), best known as the man who calculated the value of Pi. The legend that gave rise to my term for the second form of inventorship goes like this. The Greek ruler Hiero II asked Archimedes to determine whether a crown was made of pure gold or of a silver alloy. Upon stepping into a bath and seeing the level of the water rise as he lowered his body into it, Archimedes realized that a given weight of gold would displace less water than an equal weight of silver (which is less dense than gold). Excited by his discovery, Archimedes is said to have leaped from the bath and run naked through the streets, shouting, "*Eureka!*" (meaning, "I have found it!"). Eureka Inventorship *seems* to happen instantaneously, but I believe the prepared mind has already subconsciously processed a lot of information relating to a problem—or even apparently unrelated to it. Then, unpredictably, a flash of insight occurs.

Eureka Inventorship should not—really cannot—be dismissed as superficial, or not soundly based on knowledge, just because it isn't the result of a lengthy investigation in a laboratory. Don't misunderstand: I'm not saying that Eureka Inventorship precludes the need for careful research; even ideas that develop this way must later be evaluated and proven. The difference between the two kinds of inventorship is that Research Inventorship is formal, disciplined, and deliberate, whereas Eureka Inventorship is informal, apparently undisciplined, and seemingly random and accidental.

Far fewer people undertake Eureka Inventorship, but more of us could if we just practiced. Few of us will ever have the opportunity to work in a laboratory or be awarded a multimillion-dollar research grant, but we can all employ Eureka Inventorship to make our lives better and more enjoyable. Learning to do so is what this book is all about.

Extend the Boundaries of Learning

By its nature, Eureka Inventorship is free ranging; it ignores boundaries. Eureka inventors can see relationships between ideas and problems that, to others, appear unrelated. Thus, Eureka inventors often extend their thinking beyond their own fields. Many of the great innovations in history came from people who were not specialists or acknowledged "experts" in the particular area where their inventions apply. As freethinking "outsiders," they were able to see solutions that those who were closer to the problem had missed.

PERCY L. SPENCER

Walking around in one of the laboratories at the Raytheon Company shortly after the end of World War II, Percy Spencer felt something funny: a chocolate bar he had in his pocket was melting. He was standing next to a radar set's magnetron tube at that moment, and, being an electronics genius, he thought he understood what was happening and why. Quickly, he asked for some unpopped popcorn. At that location, it popped in the open air.

Before long, Raytheon had built the first microwave oven for commercial food services. It weighed 750 pounds and measured almost six feet in height. Many years passed before the technology was streamlined for popular use, but Spencer's invention is now in most modern homes throughout the world.

ALFRED C. GILBERT

While at Yale Medical School, Alfred Gilbert performed magic to earn money for his tuition. He was an outstanding student and athlete. In 1908, he took time out from his studies to compete in the fourth Olympiad in London, England, where he won the Olympic gold medal in the pole vault (thanks in part to an innovative, spikeless bamboo pole that he was the first Olympic contestant to use). He was also an entrepreneur, having manufactured a children's magic kit called the Mysto Magic Exhibition Set.

After finishing medical school in 1909, he decided to continue in the toy business and, inspired by the railroad bridges he saw while riding on a train to New York, he assembled an engineering construction kit for children. Its metal beams and joints had evenly spaced holes for nuts and bolts. This was the Gilbert Erector Set of 1913. It continues to be one of the most creative and popular toys ever designed.

ABIGAIL M. FLECK

When she was eight years old, Abbey Fleck was helping her father cook bacon and noticed how much fat needed to be drained onto paper towels after the bacon was cooked. She wondered why the slices couldn't "drip dry," and she did more than simply think about it. She designed what she called the "Makin' Bacon" system.

It is a square, one-inch-deep, microwave-safe plastic pan with three vertical supports on which the bacon cooks. The fat drains into the pan beneath. After cooking, the fat may be easily poured off without using any paper towels for draining. During cooking, one towel is placed over the top of the slices to keep the fat from spattering.

She has two patents on the "Makin' Bacon" device, which is available at many retailers and on-line.

CLARENCE BIRDSEYE

Taking time off from his studies as a biology major at Amherst College, Brooklyn-born Clarence Birdseye took a job as a naturalist for the U.S. Government. He was posted to the Arctic, where he observed the Native Americans' way of life. He was especially intrigued by how the combination of wind, low temperature, and ice froze fish almost instantly after they were caught. Even more important, he discovered that when the frozen fish were later cooked, their taste and texture seemed the same as if they had been cooked fresh.

As a scientist, Birdseye saw that the fish were frozen so fast that no ice crystals formed in their cells. The idea caught his imagination. He began to design and patent machinery and systems that would allow high-pressure, quick freezing of food far south of the Arctic. In 1924, he founded Birdseye Seafood, Inc. He also designed frozen food storage units for grocery stores and specially refrigerated railroad cars to transport the new product.

Today, most frozen food is still processed using basically the same method that Clarence Birdseye imagined when he first saw fish frozen in the Arctic.

From my personal experience among both Research and Eureka inventors, one example stands out. It struck me with particular force when I was on a panel of 15 inventors who had been inducted into the Inventors Hall of Fame. As I looked around that group, I realized that only two of us—myself and Dr. Forrest Bird—represented the Eureka contingent.

You've probably never heard of Dr. Bird, but you undoubtedly know about his great Eureka invention: a passenger's oxygen mask for use in high-altitude business jets. As these aircraft became the standard and business-jet air travel became the norm, there was a clear need for an oxygen mask to serve as an emergency source of oxygen in case the pressure inside the cabin of the plane failed for any reason.

The mask was designed to deliver very high-pressure gas at normal pressure to the passengers and crew if the cabin pressurization failed. Dr. Bird's idea was simple: Balance a large-demand valve in the mask with a tiny valve on the high-pressure oxygen tank. When the mask wearer's breathing opens the large-demand valve in the mask, it triggers the opening of the tiny supply valve of the very high-pressure oxygen tank. The ratio of the large valve, which delivers normal-pressure oxygen to the user, to the small valve, which releases oxygen from the high-pressure supply tank, is Dr. Bird's "another way." (We'll return to Dr. Bird's invention later in this chapter.)

Eureka Inventorship allows us to make full use of the knowledge and experience we accumulate throughout life. Most of us abandon this valuable knowledge because we have stopped exercising our childlike sense of wonder and our once-flexible imagination.

Seeing What's in Front of You

The tag line from *The X-Files*, one of the past decade's most popular television programs, tell us: "The Truth Is Out There." I

THINKING UNDER PRESSURE AS A FORM OF INVENTORSHIP

For almost 100 years, the tragic sinking of the supposedly unsinkable ocean liner *Titanic* has intrigued the public. Since that cold night when the ship and more than 1,000 of its passengers fell victim to an iceberg, people have speculated whether the extensive loss of live was unavoidable. While, sadly, some of this speculation has been little more than finger pointing, some is well founded.

This terrible event offers an illustration of how innovative thinking works and how "another way" might have saved lives.

According to research (reflected in the most recent film about the disaster), as a result of the collision, the forward compartments of the hull began to fill with water. Eventually, the weight of the incoming water caused the bow to begin to sink. The bulkheads forming the "watertight" compartments were not designed to extend to the deck above, so as the bow sank further, water overflowed from one compartment into the next. As one compartment after another began to fill, the bow was depressed even more, until, at an angle of approximately 34 degrees, the great ship began its rapid slide to submersion. Within an hour and 40 minutes, the unsinkable *Titanic* had vanished into the deep North Atlantic.

Like so many others, I too have speculated on other choices or emergency actions that might have been taken that night to lead to a less disastrous outcome. Why, I have always wondered, didn't the engineer tell the captain to open the seacocks in the stern of the vessel, admitting water to the stern compartments to balance the water coming in at the bow? This would have prevented the ship from reaching the fatal 34-degree angle so quickly. Probably, this alternative solution was not considered because the problem—the water infusion—was at the bow, and everyone's attention was focused there. Undoubtedly, widespread panic played a major role in the failure to think inventively in this situation, but it is interesting to recognize in retrospect that sometimes the solution to a problem is exactly at the opposite end of our focus—in this case, at the stern, not the bow, of the sinking ocean liner.

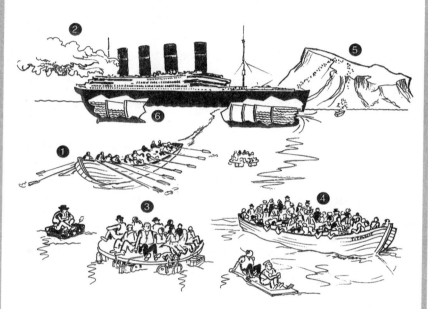

Many innovative options were available to the leadership of the *Titanic* if they had only been able to think about very different solutions to the basic problem: (1) Crews of the most fit could have rowed out to the closest vessel. (2) A signal fire could have been set on an upper deck to attract attention. (3) Rafts could have been constructed from a variety of floating materials onboard. (4) In the calm seas, the lifeboats could safely have been loaded to well above the prescribed limits. (5) The more athletic passengers and crewmembers could have climbed onto the very Iceberg that had caused the damage. (6) The stern compartments could have been slowly flooded to offset the flooded compartments at the bow, keeping the ship more level and thus afloat longer.

(continued)

THINKING UNDER PRESSURE AS A FORM OF INVENTORSHIP *(CONTINUED)*

Could more innovative thinking have saved lives on that fateful night? I believe it could have. As difficult as it may be at such drastic times, it is important to look not only in another direction, but also in *every* direction, and to explore *all* the possibilities. Under stress, the need to think around the immediate problem is often ignored, but if we train ourselves to do so, we will see many ways to deal with the problems that confront us.

I often use the *Titanic* story when I speak with groups of school children about inventorship. They usually come up with many other innovative ideas that might have saved lives on the *Titanic*. Here are a few examples:

There were two steamships in the vicinity. One was stationary, within sight, and one was several hours away—more hours than the *Titanic* stayed afloat. Prolonging the floating might have allowed those vessels to arrive on the scene in time for more passengers and crew to be saved.

The nearby ship mistook the emergency flares for fireworks at an on-board celebration. A flaming signal fire set on an upper deck would have signaled real distress.

Rafts might have been constructed by lashing together anything that would float—storage barrels, doors, mattresses, tables, and so on.

It was a calm night, so the lifeboats could have been loaded well above their stated capacity by having some passengers lie under the seats and others on top of one another.

Some of the more able-bodied people might even have stayed out of the water on the iceberg, which was still nearby.

The most able, strongest, and fastest rowers might have taken a lifeboat to the nearby ship to ask for help.

If your efforts to solve a problem are coming up empty, try looking away from the problem, in the opposite direction. Be counterintuitive. The biggest problem with the *Titanic* was *not* that the bow was filling with water. It was that *only* the bow was filling with water, thereby bringing the vessel progressively closer to an angle that would seal its doom. Paradoxically, the same thing that was causing the problem—water entering the ship—could also have eased the problem.

submit that the *solution* is out there, too. Sometimes, invention is less a process of conceiving a new solution than one of discovering a solution that's been out there all along but has never been applied to a particular problem. Let me share with you how this theory came to shape my own career in aviation.

Every morning, before I've had my first cup of coffee, I find dozens of others in the business of aviation congregating in my backyard. No, they're not airline executives or aeronautical engineers. They're birds. Ironically, we sometimes dismiss innovative but unconventional ideas as being "for the birds," but I for one am not so quick to speak disparagingly of our feathered friends. They inspired my first invention. Because of something I learned from birds, people everywhere now travel in airplanes much more safely than they did 60 years ago.

I learned that tiny hairs on their wings help birds sense the airflow over the leading edges of their wings when they fly. This sense of airflow is extremely important for human pilots, too, because the air flowing over and under the wings produces the lift that keeps planes airborne. The Stall Warning Indicator tells pilots about the airflow at the leading edge of the airplane's wings in much the same way that the tiny hairs on the birds' wings keep them informed. How? Let me explain further.

If the critical lift that keeps an airplane flying is lost, a stall occurs. Before going any further, I must comment on the word "stall." Most of us have experienced the frustration of having an automobile stall on us, and that's what most people think of when they hear the word. But that use of the word refers to an entirely different phenomenon—a shutdown of an automobile engine, which can be caused by any of several problems. In short, aircraft stall has nothing to do with engine stall.

However, another, more appropriate, automotive analogy can help to explain aircraft stall: the familiar skid. What we experience in a skid (besides the adrenaline rush of panic) is loss of control. The car no longer responds to steering and is no longer going in the direction toward which its wheels are pointing. If we can't regain control, we may "spin out" or even crash. Though the operating dimension is different, something very similar happens when an aircraft stalls. When driving a car, the act of turning the steering wheel changes the horizontal direction of the car's wheels (and hence, the car's direction) toward the right or the left. In an airplane, the primary directional control is vertical—up or down.

How is an airplane's vertical direction controlled? With its wings. In flight, the airplane's wings force the air downward, which results in lower pressure over the wings than under them. This produces a sort of suction on the wings, or what we call *lift*. Stall happens when the wings present too steep an angle to the air they are moving through. This prevents the air traveling over the top of the wings from going fast enough to rejoin the air moving under the wings. The airflow over the wings is broken, preventing them from producing the necessary lift.

In much the same way, control of a car is compromised when its tires lose contact with the road surface and begin to skid. An automobile skids because the angle at which its wheels can effectively turn has a limit, depending on the speed at which the car is traveling and the road conditions (unpaved, wet, or icy, for instance). If you exceed that limit because you want a more sharply angled turn, you'll actually get less, because your tires lose side-traction. Your car continues forward, but not on the path you intended.

As just explained, an airplane stalls when the angle at which air can flow smoothly around its wings is exceeded. (See Figure 1.2.) You want more lift, but you get less because the wings, which had been slicing nicely through the air, are now meeting the air at too great an angle. The wings are no longer

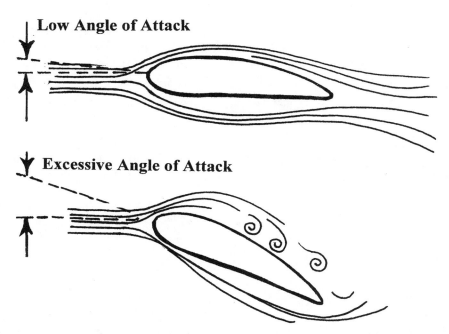

Low Angle of Attack

Excessive Angle of Attack

Figure 1.2 The wing diagrammed at the top is in normal flight with proper airflow. The diagram at the bottom shows a wing that has stalled because the disturbed airflow over the wing is not producing lift.

grooving the air around them so that the aircraft follows a predictable air track. Instead, they are directing the plane into an unintended path. And if a tall building or the ground is in the flight path when you're in a stall, the results, obviously, can be disastrous.

In the early days of aviation, before pilots understood this phenomenon, they called its effect "the Death Dive." Airplanes would go up, then would suddenly dive down. At sufficient altitude, airplanes might recover, but more often than not, they crashed. When the airplane was dropping, it might spin, corkscrew, or otherwise behave uncontrollably.

I became interested in solving the stall problem when I happened to witness a plane crash near the airport where, as a novice pilot, I regularly flew. I later learned that the pilot had maneuvered the airplane in such a way that its wings were at too steep an angle to the airflow, and the plane stalled. As I watched, it seemed to just fall out of the air.

Pursuing answers to this perplexing and dangerous problem, I asked more experienced aviators why the pilot hadn't simply decreased the angle of the wings to the airflow to maintain control. They told me that he probably had no idea he was getting into trouble, because it's very difficult to *feel* when the wings are about to stall.

From those conversations, I was able to define the problem:

The pilot must be warned in advance of a stall.

That led me to define the solution:

Design a mechanism for indicating to the pilot that the airplane is going to stall.

In a very short time, I came up with the product concept.

REPORT TO THE EDITORS

Dr. Greene fingers his stall-warning device.

Look Out! She's Stalling!
By WESLEY PRICE

DR. LEONARD M. GREENE, a young scientist with a hair-trigger curiosity, started flying in 1937 for fun. Before he soloed he had the experience of seeing a fellow student stall, crash and die. Old-timers dismissed the accident with: "Pilot error. He should have felt the stall coming. You can feel it in the seat of your pants."

It didn't sound very scientific to Doctor Greene, this seat-of-the-pants stuff. He knew that a stall affected the lifting power of the wings, not the engine: that a too-slow airplane lost its grip on the air and plunged earthward like a spent arrow. But how to avoid stalls? Instructors tried to explain to him. This nose-high attitude warns of a stall, doctor—but watch out for nose-down stalls too. Your air speed is the best guide, doctor; it's safe at seventy mph—unless you're in a tight turn. If in doubt, trust your muscle sense; just before a stall, your controls feel sloppy—that is, usually; the plane shudders—maybe; and you're light in your seat—or heavier. It all depends.

This rag bag of ifs, buts and exceptions annoyed Doctor Greene, who had been schooled in the sharp-edged certitudes of engineering, mathematics and chemistry. His flight training, begun as a hobby, gradually assumed the dimensions of a scientific inquiry. He won a private pilot's license, a commercial license and an instrument ticket. He bought an airplane and festooned the cabin with gauges. He studied aerodynamics. In time he became a professional designer of war planes and author of a learned monograph on supersonic flight, *The Attenuation Method of Compressible Flow*.

Airmen may be thankful that Doctor Greene took flying so seriously, for he has invented a gadget which takes all the guesswork out of stalls. He calls his device a Safe Flight Indicator. It detects automatically the onset of stalls and gives ample warning with a red light and a loud horn blast. It is simple, cheap and foolproof. And it may be the greatest lifesaver since invention of the parachute.

Inadvertent stalls are the most fearsome man-killers in private flying. They cause more than half of all flivver-plane crack-ups and take more than 150 lives every year. Fatalities usually occur when the airplane is low, slow and turning—as in the landing approach, the take-off climb or while the pilot is waving proudly to Nellie, look, it's me. Lest Nellie become a widow before she's a bride, Doctor Greene has begun mass production of the Safe Flight Indicator at a factory in White Plains, New York. Heart of his invention is a hinged slip of metal projecting from the leading edge of a wing, like a blackened butter knife. This closes an electrical circuit, which sets off the red light and the horn. Simple? Absolutely. Yet experts have been trying to build a stall-warner, and failing, since the first World War.

Seat-of-the-pants pilots were tested recently by Professors R. J. Hulson, of Harvard, and Floyd Dockerey, of Ohio State. Their research airplane carried five Safe Flight Indicators, minus horns. The telltale lights, visible to check pilots, were concealed from the guinea pigs. Tests showed that the average student or private pilot can't recognize the coming of a stall; that many instructors are no better than their students; and that some pilots, even experienced ones, teeter near a stall condition in their normal flying.

The moral is plain: if you fly by the seat of your pants, better let Doctor Greene electrify your rear bumper.

My idea was to put, on the leading edge of an airplane's wing, a movable vane that would sense the airflow angle over the wing in much the same way that the tiny hair sensors do on birds' wings. (See Figure 1.3.) Today, my stall warning invention (now in more sophisticated forms) is implemented universally on aircraft and has saved thousands of lives. Though the government gave me a patent for the invention in 1946, I always acknowledge the inspiration of the birds, which had been taking advantage of "my" idea for thousands of years. All I did was observe what they were doing and figure out how to adapt it to the human form of flight.

Once you tune in to this concept of inventorship, you will start to see, as I have, hundreds, even thousands, of people successfully using it as a problem-solving approach. The origin of

Figure 1.3 My invention, the Stall Warning Indicator, was a mechanical and electronic version of the natural sensory system that birds have on the leading edges of their wings.

the word "invent"—from the Latin *invenire*, meaning to come upon, find, meet with, or encounter—suggests that what we invent may, in fact, already exist in some form, and what we really do is find it. *Invent* does not mean create something out of nothing.

The inventorship idea that might make your life better in some way is probably already out there—on the wing of a bird or in a hundred other places. You only need to become observant and to open your mind to the possibilities. Then, you're likely to encounter it.

Asking "Why Not?" as a Form of Inventorship

One easy way to "exercise" Eureka Inventorship is to ask yourself, on a regular basis, "Why not?" This simple, two-word query can open the floodgates of innovation. Many of the advances we enjoy today came about because someone asked that question. There are hundreds of examples. Here are a few:

- Sprinkling clothes with water before ironing them makes the task easier and the result better. *Why not* find a way to put the water you need inside the iron?

- For as long as humans have been writing, they've been scribbling notes to themselves and others, and then losing those notes. *Why not* make notepads in which individual pages are backed with an adhesive that makes them as easy to remove as they are to apply?

- If the caffeine in coffee is detrimental to health, *why not* find a way to remove it from the beans?

- To ensure the safety of children riding in the back seat of a four-door car, *why not* devise a mechanism that prevents locks on the rear doors from being opened from the inside?

- No one can be available for phone calls at all times. *Why not* build a machine that can essentially act as a personal assistant, answering the phone, recording a message, and delivering it later?

- Serving water in drinking glasses raises the risk of transmitting germs, in addition to the potential danger from breakage. *Why not* make disposable cups out of paper?

As these examples demonstrate, asking "Why not?" doesn't necessarily lead to the invention of items of great scientific import. Inventions can be simple—in retrospect, even obvious—and still have a valuable, long-lasting, positive impact on people's lives. Post-it notepads, telephone answering machines, paper cups, childproof door locks, and decaffeinated coffee are just a few of the countless benefits we all enjoy because of people who practiced inventorship.

Adapting a Solution as a Form of Inventorship

Sometimes, inventorship used to solve one problem may lead to solutions for other problems as well. Remember Dr. Bird and his oxygen mask? That invention came to have more far-reaching implications. True to the nature of Eureka inventors, Bird expanded the possibilities of his innovation beyond his own field and, in the process, helped a great many others besides air travelers.

What he had learned while developing oxygen masks for private business jet aircraft contributed to his design of a small, portable respirator machine that helps premature infants to breathe. He adapted his original system to work in a closed incubator, rather than an enclosed plane cabin. (See Figure 1.4 on page 22.) Eventually, this life-saving machine came to be used around the globe, especially in Third World nations.

In my own career, the solution to one problem solved a later, seemingly unrelated problem that had nothing whatever to do with my work as an inventor. The initial problem was: How to develop an innovative employee plan at my company, the Safe Flight Instrument Corporation, to ensure that we could recruit, and then keep, high-quality personnel. The solution was a comprehensive—and, at the time, innovative—benefits package. (Eventually, I'm proud to say, we won honors from government officials for our enlightened personnel practices.)

The second problem arose when we had to seek a new location for our company, due to urban renewal in White Plains, New York, our first home. We had our eye on a parcel of land near the Westchester County Airport. It was ideal for an aeronautical company like ours; unfortunately, it wasn't zoned for manufacturing. We thought we were facing an uphill battle, knowing that rezoning for industrial purposes often meets with fierce resistance. We could have gone the obvious route—hired a zoning attorney and made a case based on expert testimony and the like—but we decided to see whether we could make it a win-win situation for the town and the business alike.

Our plant manager, who also served as chief of the volunteer fire department in the town where the desired property was located, approached the town board, showed the members our employee benefits manual, and then asked a simple question: "Isn't this the kind of company you want in your town?" (See Figure 1.5 on page 23.)

Needless to say, the board rezoned the land we wanted, and it has proven to be a perfect location for us. But none of us

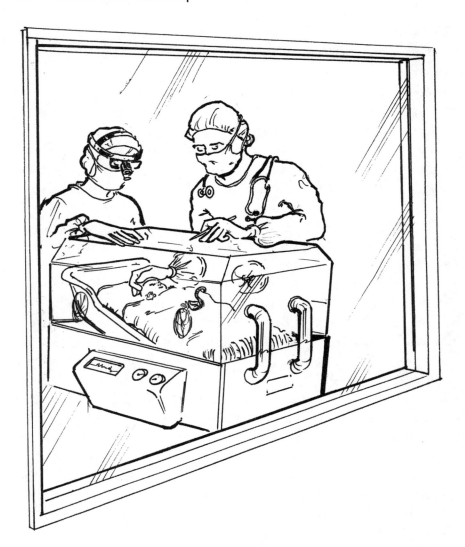

Figure 1.4   Dr. Forrest Bird's oxygen delivery system, designed for jet aircraft and modified for use in premature infant nurseries throughout the world, is a great example of Eureka Inventorship.

Figure 1.5   Solving one problem often paves the way for solving others.

would ever have imagined that by solving our employee problems we would later be able to solve a larger-scale business problem.

Getting Started!

Just starting to do something seems to keep inventorship going. Perhaps our minds are subject to a sort of Mental Law of Inertia. A body in motion tends to remain in motion; maybe a mind that has at least gotten started, however well or poorly, also tends to keep going.

In this book, I will discuss, using personal real-world examples, how you can bring inventorship into your own life. To help you do this, each chapter ends with some "inventorship points"—brief reviews of inventorship issues. I encourage you to focus on and explore them as we proceed. They will help you tap into, then develop, and eventually become comfortable experiencing the inventorship mind-set. Get started using it in your life, and you'll find it surprisingly easy to continue. You will keep solving your problems inventively because inventorship really works. And, it's a whale of a lot of fun.

Here are the inventorship points of this first chapter:

- There's a big difference between inventions and patents. People can be inventive and never take out a patent.

- Everyday experiences contain many innovative secrets, but they are revealed only to those who have developed the necessary awareness.

- Eureka Inventorship—as opposed to Research Inventorship—is imaginative and spontaneous. It ignores boundaries. Eureka inventors can see relationships between ideas and events that, to others, seem to have no connection.

- By practicing Eureka Inventorship, we can make full use of all the knowledge and experience we accumulate throughout life.

- To encounter inventive solutions, you only need to become observant and open your mind to the possibilities.

- To "exercise" your Eureka Inventorship muscles, ask yourself on a regular basis: "Why not?"

- The inventorship used to solve one problem may lead to innovative solutions for other problems as well.

2

The Age of Innovation

*I do not know what I may appear to the world, but to myself I
seem to have been only like a boy playing on the seashore, and
diverting myself in now and then finding a smoother pebble or
a prettier shell than the ordinary, whilst the great ocean of truth
lay all undiscovered before me.*

—Sir Isaac Newton (1642–1727)

By "The Age of Innovation" I don't mean the seventeenth
century, highlighted by Newton's discovery of gravity, and
Galileo's invention of the telescope and his subsequent exploration of the heavens. Nor do I mean the nineteenth century, in
which the great inventions included the steamship, railroads,
electricity, submarines, automobiles, and so much more. The
Age of Innovation I am referring to is not a historical age at all.

It is one of the ages of humankind. The Age of Innovation I refer to is childhood.

We are born with instinct but not with knowledge. Human infants learn very rapidly, however, from experience and observation, from trial and error. To watch a baby develop is to see innovation at work in its purest form—although grown-ups tend to see it as play. First there is exploration—often random and happenstance. Each successive "hit" of knowledge leads to more experiments, which in turn lead to ever broadening understanding. Beginning with their delight in the discovery of their own fingers and toes—what they taste like; how it feels to grasp objects with them; the power they confer (to throw off blankets, for example)—infants and young children take obvious joy in their expanding mastery of the world about them. To them, it's as natural as breathing; learning *is* life to them.

But compared with animals, we humans are pretty slow learners. Most other species of animal are able to make their way in the world soon after birth, whereas human children remain dependent on their parents for years. Yet we're supposedly the most intelligent species. How can we account for this paradox?

As an inventor, I see this prolonged maturing process as what enables us to achieve innovation that is denied to other creatures. Yes, a bird learns to fly, hunt worms and insects, and build a nest relatively quickly, but birds cannot devise a faster or better way to fly, a more effective way to capture food, or a better way to design a nest. We humans, however, can find new and better ways to do virtually everything we need to in life. Yes, we share a lot, physically and behaviorally, with a number of other animals, but we have a monopoly on inventorship.

What does the long learning curve have to do with being inventive? A great deal. Think about it: Eureka inventors are not always successful on their first try. Sometimes—often—

they achieve their goal only through a lengthy period of trial and error, doing something wrong hundreds, perhaps thousands, of times, until they finally succeed. We fail many times to succeed once.

Encouraging Innovation

How do we teach children to be innovative? There is a simple answer to that question: Leave them alone. Adults often view their children's experimental stage of growth as just play. That view tends to discount the value of this type of inventive activity. That is a mistake; such "play" is typically free of constraints and is therefore open to all possibilities. When the human mind is open to all possibilities, remarkable things happen and remarkable connections are made.

For example, while once visiting my grandson's kindergarten class, I asked the children to make paper airplanes. One child's aircraft, in particular, caught my attention. The little girl had cut the wings to make them ragged. When I asked her why she had done that, she replied, "Those are the feathers." This child had made the connection between the wings of birds and the wings of airplanes, and used her imagination to adapt one to the other. Such ideas epitomize the creative nature of children's play.

Parents should encourage this type of experimentation during play, taking care not to impose their own notions of right and wrong on the situation. Many of the best toys don't come from Toys "R" Us; they are created by imagination. A child's inventive play can turn his or her own hand into a bird or a plane, which, when zoomed and dipped through the air, can provide hours of entertainment.

VALUING MISTAKES

Before long, of course, the necessary process of training children begins. Unfortunately, training often incorporates a long list of don'ts. Teaching that there is only one right way to do something is the sure path to stifling creativity and passing on the concept that making mistakes is to be avoided. Instead, making mistakes is essential to innovation. Parents should do everything possible to encourage—not discourage—their children's trial-and-error tendencies, for this capability will help them to be able to tap into their inventorship to solve problems throughout their lives. Parents, therefore, must negotiate the narrow path between anarchy and repression. They must guide where necessary, without stifling creativity.

Young children naturally do things wrong on the road to learning to do things right. They try to walk, and they fall. They try to eat, and they make a mess. They try to ride a bike, and they take a tumble. Eventually, they find effective ways to master all these skills—but only after they try many ways that don't work. And in the process, they learn something more important than how to stand up on their own two feet or to get more food into their mouths than on their clothing:

They learn how to learn.

They find out that trial-and-error works. And that is the lesson of inventorship.

Childhood is very important to the development of inventorship. In the trial-and-error process of learning, children also begin to observe and evaluate the results of their efforts. From this learning stage, they discover how to look at a problem and imagine a solution.

Talk to people who have developed the art of innovative thinking, and you'll probably discover that they were encouraged to approach tasks and problems with this mind-set at a very young age—at the time when innovative thought processes come more naturally. One of the problems of the formal learning process is that adults constantly label things as "right" or "wrong" and make it clear that the wrong way is to be avoided. This closes off the opportunity to experiment and learn other ways that might lead to "an *other*" right way to do something. (In Chapter 1, "The Nature of Inventorship," I made an important distinction between another and "an other.")

Parents do not need to teach children to be innovative. They need to allow them to be inventive. That may mean permitting them to do some things that adults would define as "wrong"—as long as doing so will not put them in harm's way or harm others, of course.

HONORING PLAY

It has been said that play is children's "work." Through imaginative play, they learn how to work out relationships with others, face their fears, explore ways of interacting with the world, and find out about themselves. The word "play" implicitly discounts these activities' value. But, in its proper context—that is, in relation to a child's capabilities—play can be viewed as purposeful—comparable to adult work. If only more adults worked as inventively as youngsters play! A playful approach to complex problems often leads to the most inventive solutions.

LEAVING ROOM FOR THE IMAGINATION

One proactive way to honor child's play is to leave room for children to use their imagination—to resist the urge to give them the latest and greatest supertoy. Most of us have had the experience of fighting crowds to buy and bring home the current, most publicized toy at a gigantic toy store, only to discover that our child plays with it for an hour or a day, but then grows quickly bored with it. Or, we've watched, dismayed, as our child is more interested in the supertoy's box.

This is a clear message from children, one that wise parents pay attention to: A large part of a child's enjoyment in play activity comes from being creative and innovative. When we give a child highly detailed objects to play with, we limit the imaginative possibilities. Thus, to a child. a wooden stick may be a better "sword" than the more realistic gilt-and-enamel version from the toy store. And when he or she is tired of swordplay, that wooden stick can, in an instant, become a spirited but deeply loyal horse—perhaps an Arabian stallion or a Pinto pony. Later still, that same stick might be a flagpole planted on the polar icecap by the brave Antarctic explorer who gets there first. Tomorrow or the next day, the stick may form the beginning of a stockade or the wall of a house. That wooden stick, so simple, so abstract, has almost unlimited potential. It can be anything a child imagines it to be.

I recall happily the day my parents bought a new stove, because they gave the young Leonard Greene the old one (plus a hammer, screwdriver, and wrench). These items came with the invitation to take the thing apart in a corner of the garage. The result? Many hours of fun. (See Figure 2.1.) Years later, I gave my six-year-old grandson an old radio tape player and a couple of tools and challenged him to see how many pieces he could get out of one tape player. He learned much more about parts and tools than I could have taught him if I had taken the machine apart for him and told him what everything was and how it all functioned.

Figure 2.1 Taking apart our old stove was more fun and taught me more than any mass-produced toy or game ever could have.

If we want to foster inventiveness in our younger generation, it is also vital that we let children have things they can play with without being afraid of breaking them. And they must be allowed to be creative even if we find them annoyingly messy or disorderly. We have to remember: It's easy to clean up the mess a child makes. It's not so easy to teach him or her to be creative later in life if he or she was never permitted as a child to experiment, tinker, or dismantle—in short, to imagine.

Giving our children things to play with that encourage their imagination to soar also means that we adults need to be

ANSWERING TOUGH QUESTIONS INNOVATIVELY

In addition to giving children tangible objects to play with to encourage healthy imaginations, parents can help children understand complex concepts by being innovative in their approach to the difficult questions children inevitably ask. A friend of mine is the father of five. Years ago, his youngest daughter, Mollie, then age five, asked why she had been born last. My friend and his wife explained that they wanted her first, but her brother Peter came first. They tried again to have her, but Richard was born instead, then Diana, then Joan. But they told her they weren't going to give up until they had her. On the fifth try, they were successful, and Mollie was born. That inventive response was exactly what Mollie needed to hear.

able to think creatively. Yes, you can go into a store and buy specially designed "educational" toys for your youngster. But you probably already have around the house some highly instructive playthings around the house passing as junk.

Understanding Misbehavior

Parents should also strive to recognize and encourage the natural aptitude of children to discover and innovate even if it sometimes appears in the guise of misbehavior. Let me give you an example. A few years ago, I was invited to speak on the subject of inventiveness to a group of research scientists at the Goodyear Corporation, in Akron, Ohio. Though my talk was very well received, I have to acknowledge that I was later upstaged by a couple of inventive youngsters—two of my grandchildren.

During lunch with 20 Goodyear executives in the corporate boardroom, the children were seated along the center of the table, and were treated to hot fudge sundaes. Unlike adults, who approach such a concoction from the rich fudge on the top, these two found a way to make their sundaes seem bigger: by making them taller. With their spoons, they dug to the base of the sundae, scooping away ice cream a spoonful at a time. The excavated ice cream didn't go into their mouths initially; it was deposited on the top of the sundae. (See Figure 2.2.) You might be thinking: "That's no way for children to behave in the staid boardroom of an important corporation," and perhaps this company's top executives did disapprove as well, but they

Figure 2.2 My grandchildren took an innovative approach to dessert by building "mile-high" sundaes.

THE CRAYON EXPERIMENT

Administering discipline is a trying experience for both parent and child. I only spanked my daughter Terry once, and the occasion was memorable. What prompted the spanking was one of Terry's early experiences with thinking in "an other" way. I spanked her because, in effect, she dared me. But later, it became clear to me that she was being experimental, not defiant. (See Figure 2.3.)

Testing the limits of authority is a creative way for children to learn how to break the rules and succeed in another way.

Terry, three years old at the time, was discovered to be "expressing herself" by using her crayons on the bedroom wall. "Do that again, Terry," I warned, "and I'll give you a spank."

She immediately turned back to the wall, crayon in hand, and continued to enhance the pattern of the wallpaper. I made good on my promise and gave her a spank. Later, I asked her, "Terry, why did you continue to mark up the wall when I said I'd give you a spank if you did?" "I wanted to see," she explained quite rationally, "if you really would."

were also fascinated by what these youngsters were doing: they couldn't take their eyes off the children and the tall, thin "ice cream architecture" they were building.

My point here is that there doesn't have to be a practical application of a child's inventiveness. Messing around or misbehaving (when it is not destructive) is another form of innovative thinking. Just because not every manifestation of creativity results in a practical invention (most don't) doesn't mean these more transient innovations are pointless. In addition to being enjoyable, they contribute to the learning process.

When properly encouraged and monitored, such "misbehavior" can grow into inventorship with very useful consequences. Adults must recognize when "acting up" or "making a mess" is a genuine expression of creativity. They must then encourage children to exercise that creativity in constructive ways.

Too many adults have lost sight of what children know inherently: Thinking inventively is not only productive, it's also fun. John Spoelstra, a marketing consultant, and former president of the New Jersey Nets basketball team, contends only half-jokingly in his book, *Ice to the Eskimos: How to Market a Product Nobody Wants* (New York: HarperBusiness, 1997), that corporate research and development departments ought to be called the "fun and games" departments. "When you scrape away all the mystique of research and development," he insists, "it *is* fun and games."

Teaching Innovation by Example

So far in this chapter, I've described how adults can help children learn via their own innovative thoughts and play. But children learn by example, too, and when adults take an inventive approach to life, children incorporate that as part of their experience. I recall vividly the inventive aspect of my father's personality.

In his day, as today, businesses had secrets—confidential information such as pricing, which they wanted to *keep* confidential. Elaborate schemes were developed to hide numbers by using letters, or other numbers, to stand in for them. Before bar codes, that was the way confidentiality was secured. But most codes could easily be broken—until my father came up with one that was very different.

Instead of using the standard ciphers, he imprinted weird-looking designs on the edges of his merchandise. As a child, I remember asking him why. He told me the designs represented price figures. I pressed him to explain further because, to me, the designs didn't resemble either letters or numbers.

To create his system, my father had made an imaginative leap to an apparently unrelated area. Long before computer programmers translated all numbers and letters, and even images, into zeros and ones, my father found inspiration in Xs and Os. Not the Xs and Os that symbolized kisses and hugs and were added to signatures on letters, but the kind youngsters used when playing tic-tac-toe.

The tic-tac-toe grid, drawn when starting the game, creates nine spaces for the Xs and Os. My father observed that the shapes created by the lines framing each of these spaces were all different. When he put the numbers 1 through 9 into the spaces, creating three left-to-right rows, he had a code that, together with a 0, could represent any number. The code was the shape of the lines that formed the space in which each number appeared.

In the illustration of this code (Figure 2.3), the 1 in the upper left corner is in a space defined by lines below and to the right, so a reversed L-shaped design translates to 1. The 9

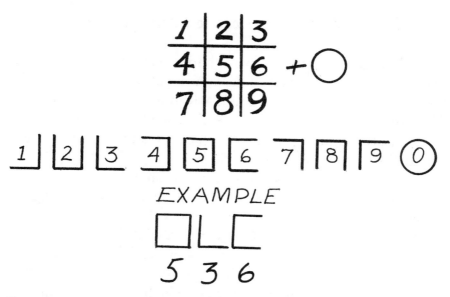

Figure 2.3 My father's brilliantly innovative "tic-tac-toe" code was used to indicate cost price; it was simple and easy to remember.

in the lower right has one vertical line and a top line that extends to the right. A full square represents the 5 in the center of the grid. Just below the 3 is the number 6. Its space is defined by lines that form a squared-off version of the letter C. This code sequence ⌐⌐□⌐ translates into a price of $19.56.

To my knowledge, no one ever figured out my father's system. To do so, they would have had to discover that the characters represented numbers. Then they would have had to discern the scheme for translating them into specific figures. Probably the last code-breaking key they would ever have thought of would have been tic-tac-toe. Ingenious, yet simple and easy to remember.

Transferring Childhood Lessons to Adult Innovation

As an adult, I too would find a better way to deal with some code—in my case, Morse code. You might be surprised when I tell you that, for 25 years, I flew without memorizing this system of communication. Whenever I needed to use it on my radio, I always had to look at a chart. Finally, I decided that I needed a better system.

Morse code, as you may know, translates each letter of the alphabet into a distinct sequence of short and long signals, which had to be committed to memory. That is, of course, unless you could find another way. Here's what I did.

First, I translated the audible short and long signals into short and long written lines. Then I arranged those lines into the shapes of the alphabet letters to which each corresponded. (The process was not unlike a child playing with matchsticks.) For example, in Morse code, the letter A is one short followed by one long signal. I envisioned the long as the right leg of a capital A and the short as the crossbar. (See Figure 2.4.) Thus, in my Morse code, A was a representation of the letter with the left leg missing. For B, which in Morse code is one long plus three shorts, I imagined the long line upright, then placed the three short lines to its right at the top, bottom, and middle—which, with a little imagination, resembled a capital B. To this day, I can recall Morse code using my system.

Many years later, I assigned the task of mastering Morse code to my Boy Scout troop when they set out to earn an advanced merit badge. I demonstrated my system to them, and tested their memories one week later. Every Scout had mastered the code.

My alphabetic system might not work for everyone, but it illustrates that, by using innovative and inventive thinking, I

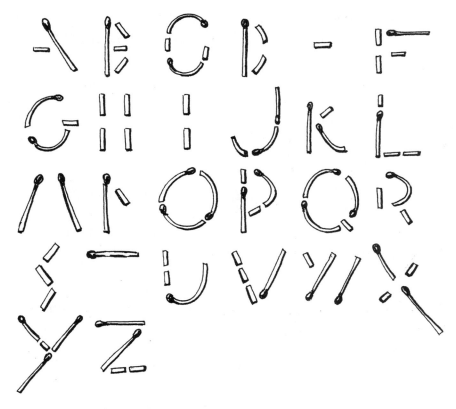

Figure 2.4 "Translating" the Morse code system of audible long and short signals into visual arrangements of long and short lines resembling the letters to which each grouping corresponded enabled me to learn Morse code easily—and forever.

was able to develop a better way to learn something that had evaded me for years.

Another image from my youth provided the impetus for inventing a medical device. As a youngster, I loved to practice shooting at a rifle range, and I remember looking down the bore of my gun, fascinated by the spiral rifling, or helix, it revealed. (Rifling spins the bullet as it travels down the barrel, giving it gyroscopic stabilization much like the spiral pass of a football.) As so often happens with images from our youth, the impression the helix made on my memory became transformed to an inventive use.

With Eureka Inventorship, you never can tell where a good idea will come from. In this case, I doubt anyone could have foreseen that a helix in a rifle would someday have a connection to the treatment of a deadly disease. I wasn't a physician or a radiologist, so I was not the logical person to invent this idea. What I did have was the imagination to make a seemingly unlikely connection. I linked a design I had first seen as a child to a completely unrelated context—the realm of medicine. Uterine cancer, though it can strike women from all walks of life, is more prevalent among women in impoverished areas, where sanitation systems, if they exist at all, are primitive. In these circumstances, cleanliness is difficult to maintain, and thus the disease is especially common in the rural areas of Third World countries.

The cancer is treated with therapeutic radiation delivered from a large, heavy, lead "safe." A small, bending pathway inside the safe contains the radiation. The treatment must be administered in a room that has lead-lined walls, to protect caregivers from the radiation. This kind of facility is rare in Third World countries. Therefore, I and one of my best friends, Dr. Norman Simon, an eminent radiologist who pioneered many procedures for treating cancer with radiation, decided to work together to find another way.

In a conventional radiation "safe," rods of radioactive material were bent so as to avoid creating a straight-line path that would enable the radiation to escape. The more the rod could be bent, the better; but too much of a bend would break the rods. I knew there had to be another way. And there was—the helix I remembered from my boyhood rifle.

I designed a radiation safe using rods shaped in a helix; this precluded the need to bend straight rods. The radioactive material was attached to the end of the corkscrew-shaped rod. It fit in the lead safe the way a corkscrew goes inside a cork. Instead of being bent, the rod could be screwed through the spiral channel formed in the safe. Because no part of the helix was straight, the risk of radiation escaping the path was eliminated. (See Figure 2.5.) The helical form also allowed the use of a solid rod. In the traditional method, the material had to be flexible enough to be bent.

This design reduced the required lead shielding by two-thirds. Unfortunately, and to my great sorrow, my friend Dr. Simon died before our experiments with the new design could

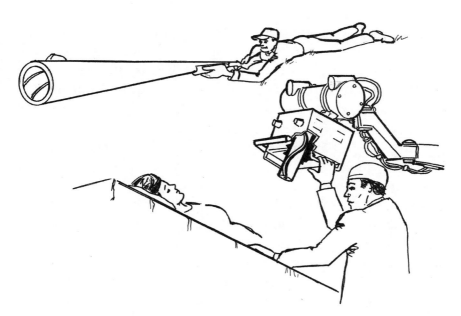

Figure 2.5 I translated a boyhood memory of the rifling in the barrel of a gun into a better way to treat uterine cancer.

be concluded, and it was never implemented. Our work does illustrate, though, that inventive ideas can come from just about anywhere, including the memory of a childhood impression. It is interesting to note other inventive applications of the helix, from the earliest rifled gun barrels to unraveling the secrets of DNA.

Accessing Random Memory to Think Inventively

Magicians at work offer a wonderful example of how linear thinking can limit us. A talented magician takes advantage of the fact that most people think linearly. All he or she has to do to work some magic is to get everyone in the audience looking away from the deception. That enables the magician to "trick" them, to perform "magic."

Let's say, for instance, that a large, sturdy-looking box is on stage, and that it features a weighty-looking, hinged top lid. Locked inside this box is the magician's assistant. The hasp of the box is closed, and a huge padlock secures it. To prove it is unbreachable, the magician strikes the locking system several times with a weighty sledgehammer. This action, of course, captures the full attention of the audience. So when the assistant subsequently emerges from the wings of the stage, they gasp in surprise.

Needless to say, the box is equipped with a second opening—one much less obvious than the top lid with its impressive lock. While the audience was riveted to the sledgehammer demonstration, the assistant opened this secret exit and "escaped" from the box, without anyone in the audience noticing her movements. (See Figure 2.6.) The point here is: The solution to a problem may often be found at some distance, mentally or physically, from where you think it will or should

Figure 2.6 The magician's skill is in redirecting the linear-thinking observers' gaze from the trick to a deception that occupies their attention.

emerge. Linear thinking often impedes the kind of breakthrough thought process that is necessary to find really inventive solutions. You have to allow your mind to operate more or less randomly—that is, in "an other" way. When our thinking is too structured, we deny ourselves access to much of what we already know.

Let me draw an analogy to the temporary storage capability in a personal computer; it is called RAM, an acronym for random access memory. Humans have a sort of random access memory, too. We have memories of, and therefore *know*, a great number of useful things. When we allow our minds to wander randomly, when we think in an apparently aimless fashion, we give ourselves access to *all* that we know, including things that may appear to have nothing to do with the problem at hand but are often the precise source of our most innovative solutions.

We are born with this ability to think randomly; too often, parental training and formal schooling stifle it. Instead, it should be nurtured. Young minds and not-so-young minds alike need the freedom to wander without limitation.

Relearning Innovation

The primary lesson of this chapter is: If you need to learn something new or to find a surefire way to remember important information, don't be afraid to think "outside the box," that is, in a distinctive—even childlike—way. Once you get into the habit of approaching problems or dilemmas in this way, you will be pleasantly surprised at how well you succeed.

Seminars and books abound that purport to teach people how to be inventive and creative. I maintain that this is not so much something we have to learn as it is something we need to *relearn*. I've said it before, but it bears repeating: We are all born imaginative. No one who watches children play games or tell stories, unhindered by adult rules, can reach any other conclusion. Too many adults, however, have become out of touch with their imaginative nature. This book seeks to reestablish that important resource.

Keep these other inventorship points in mind as we move forward:

- We fail many times to succeed once.
- Trial-and-error works, and that is the lesson of inventorship.

- Creativity and innovation form a large part of a child's enjoyment in play activity.
- Making mistakes is essential to innovation.
- A playful approach to complex problems often leads to the most inventive solutions.
- Not every manifestation of creativity results in a practical invention (most don't); but that doesn't mean these more transient innovations are pointless.
- Linear thinking often impedes the kind of breakthrough thought process that is needed to find really inventive solutions.
- Young minds and not-so-young minds alike need to range freely without limitation.

3

The Rule of Inventorship

The Wright Brothers flew right through the smokescreen of impossibility.

—CHARLES F. KETTERING (1876–1958)

The rule of inventorship is that there are no rules for engaging in inventorship. In fact, those who most successfully tap into their inventorship capabilities become adept at constructively breaking rules and at taking care not to impose rules upon themselves unnecessarily.

From kindergarten to the executive suite, we're taught to follow rules, and we're rewarded when we do so. Yet the really valuable rewards often go to people who do the exact opposite: break the rules that everyone else in their field follows or assumes to be in place. So why are we never taught the value of knowing when it is a *good* idea to break the rules? That's what this chapter seeks to remedy. Learning when to break the rules is a necessary skill for developing an inventorship mindset. Inventorship includes not limiting our choices or possibilities by adhering to irrelevant or even nonexistent rules.

The ability to consider all the possible approaches to a given situation lets us not only look at it differently, but also deal with it differently. It enables us to realize that our view is not the only view. I'm reminded of the old joke about the mother who, when asked by her son, "Mom, where did I come from?" launched into the whole story of conception, pregnancy, and birth, in terms he could understand. When she finished, she asked the boy if he had any other questions. He responded, "That's very interesting, Mom, but I still want to know where I came from, because Tommy says he comes from Buffalo."

More practically, consider for a moment how clothing fasteners have changed over the centuries. In the fourteenth century, the invention of the button was no doubt seen as a brilliant and more effective alternative to ties and other devices. (It has, after all, continued to be used hundreds of years later.) But clearly it wasn't good enough for the person who sought a better, or at least an alternative, way. And so the zipper was invented in 1926. More recently, Velcro is someone else's idea of a better way. The point here is: Each new development can spark successive inventions, proving that, in most cases, there is no one "right way" to do anything.

Using Inventive Thinking to Solve the Unsolvable

Innovative thinking sometimes enables us to explain outcomes that seem to fly in the face of what is not just a rule, but a recognized phenomenon. Let me share with you two examples of the importance of this inventorship capability in my own life.

BREAKING THE SOUND BARRIER

Flight at supersonic speeds creates damaging (and deafening) shock waves that we call "sonic boom." For years, it was thought that there was no way to overcome these effects. Nevertheless, breaking the sound barrier preoccupied aerodynamicists all over the world. As a young test pilot and research engineer, the problem intrigued me, and, with some inventive thinking, I figured out a way to solve it.

I began by adapting the way Grumman tested aircraft wings to measure airflow characteristics. Tests were often conducted on a wing that spanned a tunnel. Each end of the wing section was attached to, and supported by, the tunnel walls. I imagined having the wing go through the walls of the tunnel and stick out on either side. This arrangement wouldn't change anything being measured inside the tunnel; the portion of the wing inside the tunnel would be identical to the portion used in the usual testing. The wing would be the same size and shape throughout its length, not tapered toward the outside tip as wings usually are. In effect, we would have an endless cross section of a central portion of a normal wing. No matter where we looked along its length, its physical characteristics would be the same. (Remember, this is all theoretical. No actual wind tunnel was involved. The entire exercise took place in my imagination.)

Now, suppose that this longer wing—which could be as long as I wanted to imagine it, even several miles long—could slide through the tunnel from one side to the other. This movement wouldn't change anything going on *inside* the tunnel, because the air in the tunnel would still be striking that same identical wing configuration. The wing's sideward movement would make no change in the air pressure. In fact, someone inside the tunnel, looking at the wing and the airflow, wouldn't be able to tell that the wing was sliding sideward—that is, if the wing were clean and unblemished or unmarked.

But suppose we paint some stripes on that wing. Now, as the wing moves sideward, it's possible to see the stripes moving sideward too. The faster the wing slides, the faster the stripes move, until they're just a blur. With that side motion added to the wind motion, the wing is moving through the air at a higher speed than it was before. In fact, if the sideward speed is higher than the wind speed, the wing will be moving through the air much faster than it is when the wing is not sliding sideward through the tunnel. But no matter what we do, we haven't changed the airflow in the tunnel.

The sliding motion increases the speed of the wing going through the air, because the speed of the wing striking the air is the sum of the airspeed plus the component of the sliding speed. In other words, the wing is going through the air faster because of the sliding motion.

Perhaps a "real-world" analogy will make this easier to understand: Say you're on the beach, running directly toward a wave that is coming toward you. There's a speed at which you can reach that wave in a given time. But if you run diagonally toward that same wave, and you want to get to it in the same amount of time, you have to run faster because the diagonal distance is greater than the direct route.

The same principle holds true for the wing in flight. In my imaginary wind tunnel, as the air is coming at the wing from the forward direction, the wing is also sliding through the two holes of the tunnel in a sideways direction. The air is impinging on the wing, but the wing, moving at the same time, is also impinging on the air, so the total speed is a combination of the two effects: the wing's side (or slipping) motion, and its forward motion.

My idea was to get the wind speed in the tunnel almost up to the speed of sound and then add the wing's side motion to that speed. The total speed relative to the wing would then be greater than the speed of sound, thereby breaking the sound barrier. But—and this is the important part—none of the aerodynamics would change because the air inside the wind tunnel would not be affected in any way by the sideward motion of the wing.

In my imaginary wind tunnel, I visualized a wing moving faster than the speed of sound, but *not* being stopped by the sound barrier. The shock wave normally caused by the speed of sound would be absent because the wing's sideward movement would be angled to the air. In a roughly similar way, a snowplow that can't get through a snow bank head-on can easily slice through it if the bank is approached from an angle. This gives a hint as to how this principle could be taken out of the imaginary wind tunnel and incorporated into a real airplane: Build the diagonal into the wing itself.

Early airplanes had their wings set straight across, which meant they met the air head-on, with impinging force. To build a diagonal force into the wing, it is necessary to angle it, or sweep it back. We call this a *swept wing*. Sweeping the wing back gives us the sideward component I described in the wind tunnel test and enables the air and the wing to meet with glancing force. To achieve higher and higher supersonic speeds, we need more and more side motion, relative to the forward motion; thus, the more supersonic an airplane is, the more swept back the wing must be. Think about the wings on military jets: they are acutely swept back, resulting in a quite pointed design. A more extreme example is the Concorde. The shockwave-free supersonic airplane concept that I patented is designed to go 50 percent faster than the Concorde, and its wings are even more swept back. When the airplane is going forward at three times the speed of sound, the speed component of the wings going outward from their leading edges is less than the speed of sound. No shock wave.

No matter how fast a supersonic airplane is, its sideward speed can be as low as you want it to be, depending on how much you point sideward instead of head-on—or how far back the wings are swept. With a swept wing, if you were at the side of the aircraft, watching it go by, the leading edge would be advancing toward you at a lower speed, because it would also be expanding out sideward toward you. But it would be going *by* you much faster—literally, in a blur.

When a prizefighter hits his opponent straight on, the blow can be a knockout. With a glancing blow, the surface of the glove sweeps to the side and the impact is lessened. Similarly, the swept wing leans away from the airflow, lessening the impact of the air and making its force representative of a lesser speed. In other words, the disturbance of the air caused by the swept wing is equivalent to that for a lower speed aircraft with a square wing. In effect, the airplane says to the air, "Ha ha! You missed me!" No longer did a wing have to face directly toward where it was going; essentially, it could face in two directions at the same time. That's what enabled the breaking of the sound barrier and resulted in the design of the swept wing. (See Figure 3.1.)

My paper on "The Attenuation Theory of Compressible Airflow," written in the early 1940s, was the first to explain how the sound barrier could be broken. The concept of the

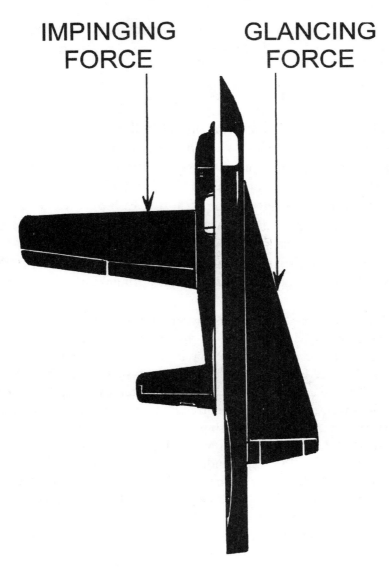

IMPINGING FORCE   **GLANCING FORCE**

Figure 3.1 The disturbance of the air caused by the glancing force of a swept wing is equivalent to that caused by the impinging force of a square wing moving at a lower speed. Thus, an aircraft with swept wings can move forward at supersonic speeds while the wings themselves are moving diagonally at subsonic speeds, and therefore causing no sonic boom.

swept wing is now basic to aerodynamics. And the swept wing is now standard on all high-performance aircraft.

GOING WITH THE FLOW

Before I discovered how "compressible flow" worked, the phenomenon could wreak havoc on pilots and their planes. How pilots experience this force is best told by my friend Corwin H. ("Corky") Meyer. At the time of this story (1944), Corky, who would later serve as president and CEO of the Grumman Gulfstream Corporation, was a test pilot at Grumman—as was I, in addition to serving as a research engineer.

Here's how Corky described the details of what happened, in an article printed in *Flight Journal*, a magazine:

> As the new Hellcat demonstration pilot, I elected to perform the high-speed dive point of 485 mph combined with the 2.5G pull-up to start the Navy load limit demonstration. At 28,000 feet on my first demonstration flight, I pushed the nose of the aircraft over to what I estimated to be a 60-degree dive angle, and totally concentrated on the buildup of airspeed versus the rapidly decreasing altitude. I was descending at about 38,000 feet per minute at full power. I estimated that I would attain 485 mph indicated airspeed just as I went through 10,000 feet, and then make the easy 2.5G pull-out.
>
> To maintain the 60-degree dive angle, I needed to continuously retrim the elevator for more nose down in order to overcome the aircraft's natural tendency to nose up with increasing speed. Just before I was going to start the simple pull-out, I noticed with amazement that the aircraft didn't require any more nose-down trim, because the nose was going down on its own. It was rapidly increasing the dive angle, without any retrimming, push force, or desire on my part!

To counter this condition, I began to pull the stick aft, but it seemed to be set in concrete. The nose-down dive angle continued to increase as I roared through 6,000 feet. It was clear that something beyond my comprehension and capabilities was directing the aircraft straight into the ground at more than 700 feet per second. I had 10 seconds to live! I yanked the throttle from full power to full closed, and pulled the stick back in both hands. After a long second or two, all hell broke loose. The airplane started buffeting violently, and pitched up to 4G. My frantic stick-pulling brought the airplane to 7G before I could release my adrenaline-assisted grip. The aircraft had bottomed out at 2,500 feet and now was zooming back up. Having had enough excitement for one day, I slowly flew back to the Grumman field in a mental fog.

I was a confused young test pilot. My mind was trying to assimilate the very frightening high-speed flight characteristics that the Hellcat had exhibited. My first thought was that a mechanic had left a wrench or screwdriver in the fuselage and that it had jammed the controls. But after a thorough inspection, nothing was found to have jammed the controls. No engineer could explain the frozen controls and automatic pitch down just as it passed 10,000 feet, and then the automatic pitch up, regaining control, as I passed through 6,000 feet.

When Corky told me what had happened to his aircraft, I could see in my mind's eye exactly what had caused it. Not only was I able to see it, I was able to explain to him, in direct, specific, and real language, how his aircraft had "broken the rules." Coincidentally, I had just completed a paper titled "The Attenuation Theory of Compressible Flow," which described Corky's predicament precisely. I explained to Corky that supersonic shock waves had developed over the top of the Hellcat's wing. This phenomenon was caused by the airflow's speeding up to supersonic velocity to go around the wing's very thick (16 percent) airfoil shape. The shock wave formation had instantly moved the normal center of lift back several feet beyond its rearward limit. These effects gave the aircraft its strong buffeting and very high stick forces (as if "set in

BREAKING THE LIGHT BARRIER

I've never read whether Einstein was aware that the formula of time, energy, and mass—by which he arrived at his theory of relativity—is the same as the formula later used to measure the properties of air at transonic air speeds. At first glance, it seems remarkable that the same small algebraic formula can be applied to both sound and light. On the other hand, why not? To some extent, light and sound are both waves (although, of course, when you get down to other theories about minimum size that can be treated as a group, light can also be defined as particles).

I believe the fact that the formula is the same may be enormously important. When, say, a thousand years from now, humans are able to make interstellar flights—assuming the human race is still around—they'll probably be doing it by breaking the light barrier. Instead of talking about speeds of Mach 1 and Mach 2, our successor humans will be traveling at the warp speeds now found only in science fiction lore. I believe we'll break the light barrier by adding another dimension to the light formula, in the same way that I was able to break the sound barrier by adding another dimension to the sound formula.

The more we learn about how things work, the simpler—and at the same time, the more mysterious—the world seems to be. There's an excitement like no other in imagining what the future may hold. Inventorship can give us that thrill.

concrete"), and kept the aircraft pitched down in spite of all efforts at recovery.

Closing the throttle had instantly put high drag on the aircraft—thus, backing it out of the shock condition, putting the center of lift back in its proper place, restoring normal stick forces, and permitting the final 7G pull-out.

Defining the compressibility flow theory and explaining the action of shock waves on airplanes in flight not only helped Corky and other test pilots to limit their test parameters so as not to encounter the phenomenon, but also proved to be of immense usefulness to designers of high-performance aircraft.

Expanding Dimensional Thinking to Tap into Inventorship

As daunted as we may be by phenomena (whether natural or manmade) that break all the rules and that we are at a loss to explain, perhaps we more commonly stymie ourselves by obeying rules of our own making that, in fact, do not exist.

Perhaps nowhere are rules so apparent as in games and sports. But I contend that even those tried-and-true pastimes lend themselves to innovative thinking—otherwise, new games would never be invented or old games improved upon. In this realm, I had the idea of making the traditional chessboard three-dimensional. (See Figure 3.2.) My variation has eight boards instead of one, enabling players to move up or down as well as forward or back. (Eventually, I patented this three-dimensional version of a centuries-old board game.)

Three-dimensionality was similarly employed in Hollywood in the 1950s, in an effort to reinvigorate attendance at the movies, whose popularity at the time was being threatened by the then-new medium of television. Today, the

Figure 3.2 Designing a three-dimensional chessboard made a very old and familiar game new and exciting.

three-dimensional viewpoint is seen everywhere—from three-dimensional maps to interactive video and virtual reality games made feasible by the wonders of computer graphics.

But it does not require a degree in any of the sciences to understand the value of transcending dimensionality to expand innovative thinking. I demonstrated this concept to one of my children years ago when she asked me to help her with a homework problem, which was given as:

> If you have a paper with eight rows of eight squares each, and the squares are alternately colored red and black, how many different pairs of adjacent red and black squares can you find?

To show my daughter that there is never just one way to approach a problem, I took a sheet of paper, and, using a ruler and pencil, drew eight rows of eight squares each. Then I instructed her to color every other one black. After she counted all the pairs, I deliberately altered her viewpoint. I rolled the paper "chessboard" into a cylinder. The result was that more different-colored squares were now adjacent to each other across the edge of the paper, so the number of pairs increased.

Given a flat piece of paper, most people would approach the problem as one in two dimensions. My question to you is: Why limit your problem-solving capabilities by restrictions that simply aren't there?

I took a similar approach to another game—in this case, a card game—in which winning depended as much on luck as on skill, a fact that annoyed me enough to want to see if I could change the outcome. Over the years, my wife and I had played thousands of hands of bridge, and it had always bothered me that the outcome of the game really didn't reflect how well we had played. Skill counts for something, of course, but the luck of the draw counts for a lot more.

I set out to find another way of playing duplicate bridge, as opposed to the more popular contract bridge, in which two teams of four players each play with the same hands. But duplicate bridge requires eight people, and, except at a party, how often are you going to find eight people who want to play bridge? I devised a way that a single team of two couples could test their skill by, in effect, playing against hundreds of other players with the same cards. Using the results of a large bridge tournament involving hundreds of participants, I compared the outcomes with the point values of the cards, then computed how much, on average, just drawing "good" cards would affect a player's results. Any results beyond that could then be attributable to skill.

In my so-called half-duplicate or handicap bridge, after each hand, players had to compensate their fellow players for the value of the advantageous cards they had received; for example, the player who drew aces and kings would pay points to the players who faced a handful of fives and threes. The number of points each card required a player to forfeit was determined by the tournament results. These adjustments neutralized the value of the high cards and made the players' skill more evident.

I had taken on this task to relieve my own frustration at an otherwise enjoyable game, but as so often happens when we use innovative thinking to solve a personal problem or dilemma, we find that it has meaning to others as well. And so it was with my half-duplicate bridge. Albert H. Morehead, author of *The New York Times'* "Contract Bridge" column, published an article entitled, "Bridge: Taking the Luck Away." It detailed my handicapping and scoring method and became of such widespread interest that it was reprinted in Canada, Russia, and many other countries. My adventure in innovative gamesmanship made bridge a more pleasurable game for many other fans of bridge.

I used the concept of expanding on dimensionality even more importantly to save my father-in-law from a major contretemps with the federal government.

Years ago, my father-in-law stored surplus vegetable oil for the U.S. government. The storage tanks were located on the New Jersey side (the west bank) of the Hudson River. On one terrible occasion, a connection broke between two tanks, and hundreds of thousands of gallons of oil spilled into the river. Though this was years before the Environmental Protection Agency had been instituted, the government was understandably upset. It had lost a lot of valuable oil, and federal representatives blamed my father-in-law's maintenance procedures for the accident.

The incident intrigued me; more to the point, I hoped to help my father-in-law out of a tight spot. I began to experiment with universal joint couplings between two cans—couplings that would move the pipes as the cans expanded and contracted. In this process, I noticed that the couplings worked in only two dimensions. If the can expanded in the wrong way,

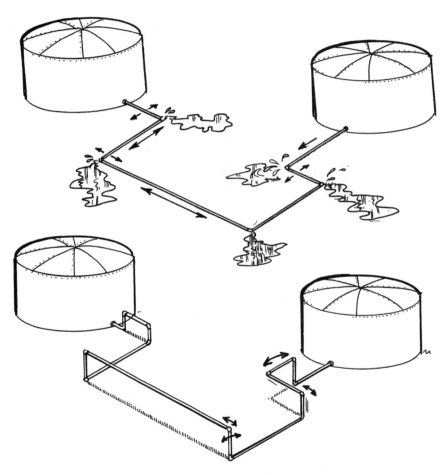

Figure 3.3 Looking at a problem in *all* its dimensions is often required before the solution appears.

the pipes could not turn, and would therefore rupture. (See Figure 3.3.)

What caused the spill, I realized, was not my father-in-law's maintenance procedures, but the specifications for the couplings. When I pointed this out to the federal authorities that had developed the specifications, the charges against my father-in-law were withdrawn.

To solve problems with inventorship, you often have to add a dimension that may not be apparent at the outset.

Thinking Innovatively to Overcome Barriers

Truly innovative thinking crosses all boundaries and goes beyond what are conventionally considered invention "disciplines," such as science and engineering. Inventive, innovative thinking is essential to those who would succeed in politics and negotiation. Innovative thinking can help all of us to navigate the often-rough waters of our relationships, both professional and personal. Unless we all learn to develop an inventive approach to interactions, we will have a more difficult road to travel to success.

OPENING COMMUNICATIONS

During his long and brilliant career, Walter Cronkite came to exemplify the capability to think innovatively. I am privileged to be his friend, and I have watched with pride as he expanded his role of journalist into the world of politics, simply by being unafraid to think innovatively on his feet. I recall his role in bringing the President of Egypt and the Prime Minister of Israel together in 1977, a time of extreme tension between those two countries.

That year, in an interview with then President of Egypt, Anwar el-Sadat, Walter asked Sadat under what conditions he would consider going to Israel to talk with Menachem Begin, then Prime Minister of Israel. Sadat's list of conditions was not surprising to Walter. Nevertheless, like the good reporter he was, he asked for clarification: "Those are your conditions for going to Israel?" "No, Walter," Sadat replied, "those are my conditions for peace."

So, again, Walter asked under what conditions he would agree to travel to Israel. Sadat said, simply, that all he needed was an invitation. Cronkite pushed the envelope: "How soon could you go?"

"At the earliest time possible," Sadat replied.

"Within a week?" Cronkite asked.

"You can say that, yes," Sadat said.

Wasting no time, Walter arranged to interview Prime Minister Begin in Tel Aviv, via satellite. During that interview, Walter of course told Begin that Sadat would be willing to visit Israel if he were invited. Begin replied, "Tell him he's got an invitation."

Five days later, Sadat flew to Israel. The rest is history.

Walter Cronkite, like so many—perhaps all—successful people, thought quickly and without regard to rules or boundaries when confronted with an opportunity, and he was able to make the most of it. That kind of thinking is a hallmark of inventorship, and it is a capability we all can develop.

TRANSCENDING LIMITATIONS

Many of us have suffered the frustration of being categorized, and thus restricted, by our professional titles and experience, by our educational background, even by our physical appearance. Perhaps we want to change jobs, or switch careers entirely, but we meet a wall of resistance. Rejecting that narrow thinking is one of the keys to successful inventorship. We must also be willing to cross the boundaries that define different areas of knowledge and experience, which, if left intact, stunt our intellectual and emotional growth and limit our enjoyment of life.

The arts and the sciences, for example, are viewed by many as having very little in common. But, for people who choose to break through these perceived barriers and engage in inventorship thinking, the rewards can be great. A few years ago, the Hudson River Museum, in Yonkers, New York, presented an exhibit entitled "The Visual Art of Music." It featured paintings by the late Ruben Varga. Two things made this exhibit unusual. First, Mr. Varga was blind; he had lost his sight when a land mine exploded near him during World War II. Second, Mr. Varga was a violinist and composer, not a painter. My organization, the Institute for Socioeconomic Studies, had commissioned him to produce more than 50 paintings, all inspired by a medium Mr. Varga knew well: music.

The exhibition was the end result of my fascination with correspondences between the senses—in this case, between seeing and hearing, or between color and tonality. I asked Mr. Varga to employ a formula I had developed years ago, in which musical notes are represented by colors associated with their frequencies. For example, the frequency of a musical tone elevated 37 octaves corresponds to a hue of light. The spectrum of visible light begins, at its lower frequencies, with the invisible infrared (literally, "under" red), which can be felt as heat, but cannot be seen by the unaided human eye. As the frequency increases, colors are revealed: red, orange, yellow, green, blue, indigo, violet, and, finally, the invisible ultraviolet. Thus, the note middle A (frequency 440) can be shown as mathematically related to the color we perceive as blue (frequency 37 octaves above 440).

Because musical works are artistically arranged patterns of notes, and paintings are artistically arranged patterns of colors, I hypothesized that it must be possible to *paint* a musical composition, thereby translating one art form into another. That's what Mr. Varga did. He used 12 acrylic paint colors to visually interpret how various composers had used 12 different notes in their musical compositions. He might portray long, slow notes as broad-brush strokes, for example, or livelier passages as a sequence of small circles. (See Figure 3.4.) If you haven't *seen* Paganini's Sonata in E Minor, you don't know what you've missed!

Categorizing fields of study or areas of interest—such as mathematics, music, engineering, and art—is without doubt useful for organizational purposes, but we must never let such structures impose artificial constraints on our thinking. We must learn to regard the lines drawn between different disciplines as boundaries, not barriers.

Consider the boundaries that modern computers have crossed—and continue to cross every day. Though the term "computer" reflects their origin, computers do a lot more than compute. They can convey—and create—music, art, literature, and more, to the entire world, in addition to serving as useful tools in all fields of study. They can do this because, in the world of computers, everything—whether a picture, a melody, a novel, or a mathematical model—is simply *information*, and information can be taken to its lowest common denominator: a pattern of zeros and ones. The human mind, perhaps the greatest computer there is, is capable of the same kinds of "calculations," enabling us to make connections between seemingly disparate entities. Remember, imagination is unlimited; do not impose restrictions on it unnecessarily.

Here are a few other examples from my own life that have taught me to believe in the limitless power of the mind.

Figure 3.4   While I assisted with the paint pots, the late musician and artist Ruben Varga painted from music, matching musical notes to their associated colors.

An inventive mind is at work even while at rest. As I explained in a previous chapter, so-called daydreaming is often an avenue along which our innovative thoughts travel. Likewise, while we sleep, our minds are at work producing dreams. If we learn to tap into this great source of creativity, it can offer us insight into, say, a problem that seemed unsolvable during the day, or bring to the surface a remarkable discovery that had lain buried in the recesses of our waking mind.

I often find that, when I awake in the morning, a solution or an idea that had escaped me the day before is clearly in mind. I have trained myself to quickly jot down on a notepad just enough of the waking thought to remind myself of it later, when I can develop it more fully. Sometimes I'll even awake in the middle of the night with an idea or two, and I jot those down, too. That's how I came to write two previous books. Each started as a simple collection of notes, some written in the middle of the night on my bedside notepad. (See Figure 3.5.) I like to think that I've been writing books in my sleep. The lesson

Figure 3.5 While your body is at rest, your mind is creatively working. Keep a notepad next to your bed to capture thoughts and ideas that come in your dreams.

here is: Don't turn off inventorship when you turn out the lights at the end of the day.

Doing something backward is sometimes the way to move forward. I learned the value of this way of thinking during a sailing trip my wife and I took, from New York to Bermuda. On the second day out, the wind picked up, making for rough and unsafe sailing. The windswept sea sent waves pouring over the deck. Below, in the cabin, the two of us were being thrown from one side of the boat to the other. If either of us went topside, I was afraid we would be swept overboard. Moreover, the conditions made it impossible for me to be at the helm.

If ever there was an occasion for innovative thinking, this was it. I realized that I had to find a way to make the boat steer itself on a steady course, to prevent the waves from pushing the craft over on its side.

As you may know, the direction of a sailboat is determined by three criteria: the direction of the wind, the angle of the sails, and the position of the rudder. To get the boat to self-correct its course, I lashed the mizzen (the sail nearest the stern or back of the boat) at an angle larger than that of the other (forward) sails, so that the wind would strike the wrong, or "front" side of the sail. I then lashed the wheel to keep the rudder opposed to the force of the wind on the mizzen. The "backward" mizzen sail countered the action of the other sails, always keeping the boat at a balanced angle between the sail and rudder forces. The boat stayed on course as if it had automatic steering. (See Figure 3.6.)

My point here is: When faced with a problem that seems to have no solution, don't be afraid to consider going opposite to the normal order of things. Put the cart before the horse. Eat dessert first. Take one step backward, if it means moving two steps forward.

The impossible may be possible. What may seem impossible at first glance can often be, upon deeper inspection, a rich

Figure 3.6   Doing something backward is sometimes the way to move forward, as I found out by lashing the mizzen sail opposite to the other sails to provide a simple self-steering function.

source of innovation. Think of something that is self-evidently "impossible," and then see whether you can find a way to do it. Here's an example of how I engaged in this mind-expanding exercise.

I wanted to prove that if you know what you're doing, you don't always have to see where you're going. This idea came to me as I thought about how difficult it was for skywriters to write in script, rather than in the more common block capital letters. I decided that it would be an even greater challenge to attempt it "flying blind." I conducted my experiment in a Link trainer (an early flight-simulator) on the ground rather than in the air. Before I got into the simulator's cockpit, I considered the various flying skills I would need in order to "write my name in smoke" while using only the simple instruments available in the simulator. (See Figure 3.7.) I had a compass, a rate-of-turn indicator, an airspeed indicator, and a watch. I then plotted—on drafting paper, in script—my first name, L-e-o-n-a-r-d. For each of the precision maneuvers, I noted the rate of turn, compass direction, and time required to "write" each letter as part of the continuous movement of the simulated flight.

Though the simulator doesn't blow any smoke, it does record flight path as a plot on paper, so I knew that the trace of my first name would be part of the recorded movements of my "plane."

Figure 3.7 Attempting the "impossible" is an inventorship challenge that can expand your mind.

To accomplish this feat with only those very basic instruments required a great deal of blind flying skill, and this was the first time it had ever been accomplished. It was regarded as so remarkable that the story was published in an aviation magazine. The fact that no one believed it could be done was the very thing that made me want to do it. To the innovative mind, the impossible just takes a little longer!

Learning That "Wrong" May Be "Right"

In life, it is true, there are a number of immutable "rights" and "wrongs," but they primarily have to do with ethical issues. We challenge those at our peril. I'm suggesting, in this chapter, that we learn to challenge those rights and wrongs that restrict our thinking, our imaginings, the rules and restrictions imposed by our own or someone else's fears or beliefs. The message of this chapter is: When faced with a problem, or while developing an idea, don't be afraid to consider opposing the normal order of things. Put the cart before the horse. Eat dessert first. Go backward if it means moving forward later. The "wrong" way may just turn out to be the right way. Remember:

- The rule of inventorship is that there are no rules.
- We stymie ourselves by obeying rules of our own making—rules that, in fact, do not exist.
- Truly innovative thinking crosses all boundaries.
- An inventive mind is at work even while at rest.
- Challenging the "impossible" can often be a rich source of innovation.

4

Defining the Problem

We learn wisdom from failure much more than from success. We often discover what will do by finding out what will not do; and probably he who never made a mistake never made a discovery.

—SAMUEL SMILES (1816–1904)

Some problems don't even appear to be problems, because they already have an accepted solution. Inventorship starts with the assumption that there is always another, sometimes better, way. This applies as much to problems that ostensibly have been "solved" as it does to problems for which an initial solution is still undiscovered. Practicing inventorship can also mean recognizing that the solution to one problem, if adapted, can become the solution to another, unrelated problem.

Recognizing a Problem Where None Seems to Exist

To succeed at inventorship, we must understand that the existence of a solution does not imply that there is no longer a problem to be solved. It may simply mean that a single, orthodox way of dealing with a situation has existed for years and has never been questioned.

To illustrate, let me tell you about the time I received a grade of better than 100 percent on a college examination. That's right, *better* than 100 percent. Impossible, you say?

I loved math in college, and it came naturally to me. I have a poor memory for rote matter, including math formulas. Given a problem, I'd usually be able to figure it out on my own, without remembering the textbook. But once, I did get an answer wrong on a test, and that wrong answer got me my unprecedented grade. The problem had to do with how many triangles it was possible to form by extending lines through points on a circle.

Let me tell you, that wrong answer really bothered me. I checked the textbook and found that the correct answer given was a figure I had come up with while taking the exam, but had rejected as incomplete. I had gone on to subtract the "lost triangles" formed by the lines that met in the points, which gave me a lower number than the textbook answer. (See Figure 4.1.)

I explained my reasoning to the professor. "The textbook is wrong," was his honest conclusion, "and your answer is right." But now he had a problem. The test had been graded on a curve, and because I had the highest number of correct answers, I had received the grade of 100 percent. Adding one more correct answer would make my score go up by 14 percent, but my grade was already 100. What could he do? He recorded my official grade on that examination as 114 percent.

Figure 4.1 Even textbooks can be wrong, yet very few students think to question them.

Another example of defining a problem that supposedly didn't exist presented itself to me at a club I belonged to. At this club were two side-by-side tennis courts that all of the members refused to use. These courts had come to be regarded as "unplayable." No one was really sure why. These were regulation courts, reportedly identical to all other courts at the club. A tennis court, as you may know, is twice as long as it is wide, so two of them can be built on a square piece of land. That fact and a contractor's mistake, I soon discovered, had caused the problem with the courts at my club. These were hard-surfaced courts, which meant they had to be capable of draining water naturally. This capability is achieved by sloping the surface downward from the high side of the court so that moisture runs off. One side of the court is then four to six inches higher than the other. When two courts are side by side, the high crown should be between them.

The first contractor had sloped the surface correctly, but a second contractor made the mistake that caused the problem no one could see. Instead of installing two side-by-side courts facing north and south, he had oriented them east and west. The highest part of the hard surface was then directly under the two nets, instead of in the nonplayable area between the two courts. Both east-west nets were as much as six inches higher than they should have been, because they were installed over the highest section of each court! "Turn the courts 90 degrees, move the nets, and restripe," I told the club managers. "Your problem will be solved." (See Figure 4.2.) They did, and it was. My reward for defining—then solving—the problem was lifetime playing privileges at the club.

When something isn't working for you, and you can't figure out why, try turning it on its side—or upside down, inside out, or backward—and see what happens.

TAKING A DIFFERENT TACK

Using inventorship to define a problem sometimes involves turning something around, as in the tennis court example. At other times, it may require that you yourself change direction. I'll use the metaphor of sailing to demonstrate what I mean.

Every sailor knows that you can't sail directly into the wind. Sailing downwind, with the wind at your back, the sails

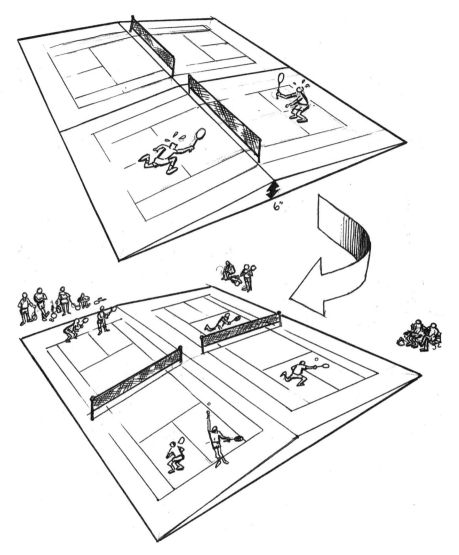

Figure 4.2 Sometimes a problem must be turned around to yield to the innovative analysis that will fix it.

fill with the breeze, and your boat slices effortlessly through the water. Conversely, sailing upwind, with the wind directly in your face, the sails "luff," or flutter, limply and ineffectually. If there were no way to move against the wind direction, sailboats would be one-way transportation, so a way was found to "cheat" the wind. Sailors call it "beating" or "tacking to windward"; landlubbers might describe it as zigzagging into the wind. To tack, a boat must have a keel to present an edge to the surrounding water and help keep the boat pointed in the desired direction.

This concept is similar to skiing or skating uphill. If your skate blade is not sharp, or if the edges of your skis aren't angled properly, you just can't do it. And you certainly can't do it head-on. On skis, you have to throw your weight from side to side, and sort of sashay up a hill. On skates, you slide your right leg out to the right at about a 20-degree angle; then, shifting your weight, slide your left leg out to the left in the same way, so that instead of attempting to move straight ahead (and making no forward progress to speak of), you're sliding forward in a series of self-correcting diagonals and, at the same time, moving right along toward your goal.

To sail into the wind, the keel and the sails work together. As in skating and skiing, the motion is not absolutely straightforward; rather, it proceeds in a series of diagonals, moving forward obliquely. An intriguing aspect of tacking is that the wind blowing *against* the boat actually gets it where it wants to go.

In ancient times, the Phoenicians knew about keels and tacking, and their swift, upwind-capable boats were the terror of the Mediterranean. Except for their fast, maneuverable craft (and, of course, galleys, which were rowed, not sailed), all the boats plying the lucrative Mediterranean trade routes had to go with the wind. And here we come to a very important fact. The wind blows counterclockwise around the Mediterranean.

(If it blew clockwise, we might have to call our culture "eastern civilization" instead of "western.") Egyptian cotton and grain went to Greece; Greek art went to Italy and Sicily; the Roman war machine went to Carthage; and Carthage shipped elephants to Egypt. The Phoenicians, who could sail upwind, became both the preeminent traders of their day (going as far afield as Cornwall, for instance, to bring back cargoes of the tin that was mined there) and the most feared pirates, because they could overtake and outrun everything else on the water.

Somehow, during the period known as the European Dark Ages, this knowledge of keels and tacking was lost. Remarkably, modern sailors fought all the decisive battles and carried on all their sea trade throughout the sixteenth, seventeenth, and eighteenth centuries by sailing downwind. Without keels, they were able to make a little headway by what was called "wearing," which induced mostly sideways motion.

The rediscovery—or reinvention, if you will—of the keel was one of the most important events of the nineteenth century. It determined the fate of nations—specifically, the relative success of their wealth and commerce. The first boats built in this "new" way (in the Chesapeake Bay area of the young United States) were able to change direction so quickly, and clipped along at such a racy pace that they were given the name "clipper." Thus began the heyday of the American clipper ship, which helped the United States win the Revolution and gave us sovereignty of the seas for a century or more.

A lot of problems in life are analogous to sailing into the wind. Likewise, a number of them can be solved by a sort of intellectual "tacking." If we can't find an inventive solution when we look straight ahead, maybe we can do it by looking just a little to one side or the other. The point is, there's almost always another way to get where you want to go.

TAKING A TWO-STEP APPROACH TO PROBLEM SOLVING

While we're still in the water, so to speak, let's examine a situation that calls for a two-step method of inventorship. Falling off a boat into the water is very serious, whether the "man" overboard is a strong swimmer or not. In fact, this is actually one of the leading causes of marine fatalities. Why? Because in just 30 seconds, even a slow-moving sailboat can be a football-field-length away from the overboard passenger. Add a few

Figure 4.3   To solve an apparently difficult problem, break it into increments that are more easily solved. The indirect approach often works best.

waves, and the person in the water may be invisible to rescuers in the boat, especially at night.

The dilemma can get more complicated. Even if the boat reaches the person overboard, or the person manages to swim to the boat, he or she must have a lot of athletic ability to climb unaided out of the water and into a boat. A man onboard might be able to pull a woman or child back in, but a man in the water might have a tough time getting back aboard unless he's unusually fit, especially if the surrounding waves are fierce.

Inventorship to the rescue. I outfitted my own boat with a small inflatable raft that ejected from a case in the hull. A person in the water can climb into a raft easily, and, from a raft, it's easier to get back into a boat. (See Figure 4.3.)

Sometimes a problem must be solved indirectly. Instead of one step, you may need two. If you can't come up with any way to get from A to B, maybe you *can* get from A to C and then easily move back from C to B.

Solving New Problems by Adapting Old Solutions

Inventorship doesn't always mean coming up with a new "thing." It can just as easily involve inventing a new use for a solution to another problem. Here are a couple of examples from my own implementation of this form of inventorship.

MEASURING DISTANCE

Architects use distance-measuring wheels to calculate distances on a building site. As one of these wheels rolls over a certain distance, the number of revolutions is recorded on a counter; which indicates the distance traveled. It seemed to me that this simple device had broader applications—specifically, to help golfers track their progress. By putting a similar counter on one of the wheels of a golf cart, a golfer could reset the counter to zero, drive the ball, follow the ball to where it came to rest, and then look at the counter. The golfer would know instantly how far he or she had hit the ball, and simple subtraction would then indicate the remaining distance to the hole. (See Figure 4.4.)

TEACHING BALANCE

Children's two-wheel bicycles often have training wheels; they keep bicycles from falling over while the children are learning to establish their balance. But, in fact, this is a faulty premise. The training wheels prevent the child from knowing when he or she is off-balance; the bicycle stays upright, no matter what. When several of my young children were learning to ride, it seemed safer to have a bike that would teach the maneuvers that destabilize the bike and the corrective actions that reestablish a rider's balance.

My innovation was an adaptation of the old bicycle-built-for-two, the tandem. I built a tandem bicycle with a difference: In front, it was a child-size bike; in the rear, it was adult-size. Both sets of handlebars could steer the bicycle. Because it had no training wheels, it was a real two-wheeler. Here's how it worked: The adult was, quite literally, the backseat driver. After the adult got the bike in motion, the child would take over the steering, to get the actual—initially unstable—sensation of balancing on just two wheels. When the child made a mistake, the adult could make the correction necessary to prevent the child from falling and getting hurt. (See Figure 4.5.)

The two sets of handlebars were cabled together, so when the adult made corrections, the child could "feel" the corrective movement through the smaller handlebars. The child learned

Figure 4.4 Innovations often come from finding a new use for an existing device—for example, adapting an architect's distance-measuring wheel to count off golf yardage.

the right moves without falling or being artificially protected by training wheels. In time, of course, the young cyclist would do more of the correcting and the adult would do less. The backseat driver would eventually become superfluous. I built one of these for teaching my own young children to ride. Later, my oldest son took it to his school and, using it there,

Figure 4.5 Modifying the tandem bicycle concept to be more instructive for the learning bicyclist resulted in an innovative new machine.

successfully taught many other youngsters to ride. With this in-novation, the would-be rider had a safer learning experience, and both novice and teacher had a lot of fun in the process. Bicycle riding without tears!

TAKING THE PATH LESS TRAVELED

It's said that you have to "know where you're going, to get there," but no one says there's one best route to take. The two examples in this section are intended to challenge you to con-sider other approaches for getting where you're going in life.

I live in an area where it's plain to see that the developers had a fancy for "ye olde English" names. Our streets have such monikers as Heathcote, Hickory, and Sherbrooke. When visitors need directions to our home, this system is problematic. It always involves the driver's counting the number of stop signs until a turnoff from an exit that has an unrelated name. Moreover, some of the streets aren't through-streets, so the driver can't determine how many blocks to go before turning again. As you can imagine, this scenario has resulted in numerous delayed dinner guests and misdirected parcels over the years. So I adapted an idea from my sailing experience. For hundreds of years, voyagers have relied on the system of latitudes and longitudes to get where they are going. Today, we depend on this system even more, because of a recent invention that makes it much simpler to do so: the Global Positioning System (GPS). GPS's capabilities go beyond those of its seaworthy predecessor. Ground soldiers use handheld GPS units; drivers depend on GPS-equipped automobiles to give them driving directions; and GPS can track truck drivers from coast to coast.

Why not, I thought, put these capabilities to work for a more personal purpose—to make it easier for friends and messengers to find us, and, more important, for firefighters, EMS, and police to come directly to our house in an emergency? Why not adapt GPS to bring order to a rather chaotic and faulty system? As you've no doubt noticed by now, one of my mantras regarding inventorship is that it always pays to take another look at how things work, no matter how long they've been around. Just because something has existed for a long time doesn't mean it has to stay that way. And just because something has been used successfully in one area doesn't mean it has to be limited to that accustomed use.

If latitude and longitude calculation can bring people safely across the ocean, or to a mooring at the next town's harbor, it can also bring people, emergency services, or packages, safely and on time, to the correct front door. What I found really exciting about this idea is that the latitude-longitude grid is capable not only of identifying a residence, but also of replacing a present telephone number and address. Just your name and the two-digit endings of the latitude and longitude of your primary residence would be enough to locate and contact you—and to differentiate you from any other John or Joan Q. Public.

Here's how my idea (which is patented, but hasn't yet been implemented) is intended to work. Suppose your latitude-longitude endings give you the four-digit combination 2143. You would simply tell friends you've invited to dinner, "Go up the parkway until the GPS in the car reads 21, then hang a right. Continue until the remainder reads 43, look for the house number 2143, and you'll find yourself at my house, which we used to call '26A Ye Olde Tavern Road.'" If your friends should be running late—not, of course, because they got lost, but because of heavy traffic—they'll simply push a button in the car, which will then automatically dial Rosedale 2143 and connect them to you. The local telephone book for the hypothetical town of Rosedale would contain, say, 5,000 names (each with its four-digit number) but no addresses. They wouldn't be needed, because your GPS ID number would function as your address as well as your phone number. Moreover, if, while your friends were at your house for dinner, a fire broke out, your automatic alarm system would call up the fire department, indicating the number 2143, and the firefighters would know immediately exactly where to go. (Note: The concept could easily be expanded globally. A ten-digit alphanumeric combination could locate any spot in the entire world to a six-meter degree of accuracy. A lesser expansion of the combination would make it practical for apartment buildings, multistory office complexes, and so on.) This idea, while patented, is far from a reality at this time. It's another example of tomorrow's use for today's technology. Inventorship at work! In a mere quarter of a century, who knows? My lat-long system may replace postal zip codes and telephone area codes, which were "invented" only in the past 30 or 40 years.

My second story about getting where you're going involves navigating by radio, but not quite in the way you might think. In a previous chapter, I mentioned the stormy sailing trip my wife and I once made from New York to Bermuda. It isn't easy to access Bermuda by water; the seas around the island are reef-studded and largely impassable. There is only one safe entrance.

From the direction in which we were making our approach, there was only one way to safely reach the entrance to the harbor: Sail beyond the eastern end of the island, outside of the shoals, then turn around 180 degrees and come in the "back door." That route requires an extra half-day of sailing. We were exhausted after the five-day storm we had just been through, but we had no choice.

To execute this move, we had to avoid the shoals; but an earlier engine failure had knocked out our electrical power, and we had no working electronic navigation equipment. All we could do, it seemed, was use the compass to aim the boat in the general direction we thought was right, and hope for the best. But hoping was not enough to guarantee our safe passage past the rocky shoals.

I had a portable radio on board—a household radio, not a ship's radio. I could pick up a station in Bermuda, but when I rotated the radio, the sound died out. I knew that when I lost the sound, the radio's antenna was at a right angle to the source of the broadcast since the least area of the antenna was exposed to the signal. That told me which way Bermuda was, but, to avoid the shoals, it was the way our boat should *not* be headed.

To conserve battery power, I had to keep the radio off most of the time. Every two hours, I'd turn it on again, rotate it to

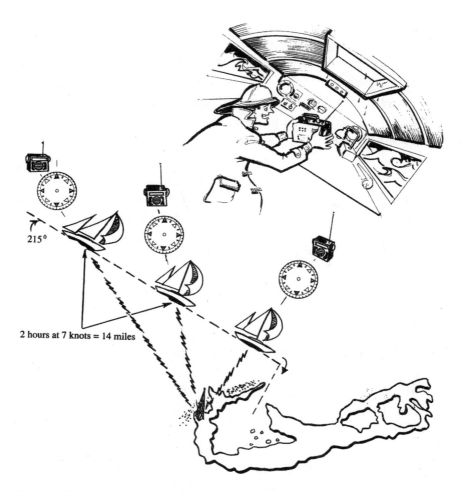

Figure 4.6 Sometimes a crisis is the mother of invention. As we approached Bermuda, a household portable radio became a direction finder in our emergency situation.

give me a fix on where we were in relation to the island, and change course based on the new information. We avoided the reefs and made it safely to Bermuda using an ordinary portable radio for which we had found a new, and possibly lifesaving, use. (See Figure 4.6.)

TURNING THE TABLES TO SOLVE A PROBLEM

One inventive way of solving a problem is to turn the tables—that is, give the problem back to its source. Let me give you an example we all can relate to.

Telemarketers are notorious for calling just when you've tucked in your napkin and have started eating a nice, hot dinner. We all know this is deliberate. They want to catch us when we're home; and, from their point of view, dinner hour is perfect timing. No one disputes that they have a difficult job, but the fact is, those calls are a terrible nuisance, if not an intrusion.

Here's another idea I've patented. I believe it could solve the telemarketer problem for all of us: We could set our telephones to filter out unwanted calls by initiating a mechanical answering process that would first ask whether the call is a personal or a marketing call. If the latter, the message would inform the caller that, if the call is accepted, a surcharge will be automatically assessed to the calling party. That way, if companies are determined to give us their marketing pitch, they will have to pay *us!* This should be an attractive idea to the telephone companies as well, because they would make money in billings. And the rest of us will all be able to enjoy our dinners in peace. If you're wondering how you can sign up for this service, I'm afraid you'll have to wait a while. The idea hasn't yet caught on, but it is being actively marketed, and I wouldn't have applied for—or received—a patent on it if it didn't have potential.

SPEEDING UP THE AGING PROCESS

I know you're thinking, "Who would want to speed up the aging process? We're all trying to slow it down." But there are many things that improve with age—for example, wine and certain other spirits. Did you know that it is possible to make certain materials age more rapidly by bombarding them with ultrasound? (Perhaps parents of teenagers suspect as much when they feel themselves getting older whenever their kids turn "their" music on full blast.) At 30,000 cycles per second, ultrasound is above the range of human hearing. For most of us, this would probably be useless information. For the inventive person with a taste for the good life, it might not be.

If you like 20-year-old Scotch or other aged wine or spirits, you know you have to pay a fancy price for those extra years. Or, you can practice inventorship by adapting a mechanism designed for a different use entirely. I bought some inexpensive Scotch whiskey and put it in front of an ultrasonic loudspeaker. I came back 30 days later with an empty glass in my hand, and filled it with the "aged" whiskey. (See Figure 4.7.) To my palate, the experiment was a success.

Perhaps 1985 was, as they say, a very good year for Bordeaux, but if you have access to the right equipment and a little imagination, you might make any year just as good.

IGNORING THE AGING PROCESS

Another way to be inventive is to ignore conventional wisdom that, plain and simple, makes no sense and causes more problems than it solves. Sometimes inventorship is just common sense. Take, for example, the subject of retirement.

Figure 4.7   Ultrasound makes almost any material age more rapidly.

In many corporate circles, an employee is a valuable and productive asset to his or her company until the day before his or her 60th or 65th birthday. The next day, he or she is suitable only to be put out to pasture. As they say, what a difference a day makes! It's nonsense, of course; but this nonsense has been codified for years in mandatory retirement rules.

IBM imposed mandatory retirement at age 60 on its executive employees. I recognized that IBM's loss could be my gain. In an ad in the local paper near Big Blue's headquarters in Armonk, New York, I invited mature executives to consider working for my very much smaller company. The ad read: "We evaluate your abilities instead of your birthdays." (The ad—which, by the way, did not name IBM—prompted a story in *The New York Times*.) (See Figure 4.8.)

Figure 4.8 An innovative ad challenged the conventional management wisdom that effectiveness in the workplace ends on a certain birthday. The ad attracted to my company some of its most productive employees.

As a result of the ad, we received a number of phone calls from IBM employees who wanted to apply. We also received a call from IBM management. The vice president for public relations was not amused; I should have called them first, he fumed. I replied that I was sorry he was upset and that in the future I certainly would call them before I ran any other advertisements—adding that, of course, I would expect IBM to show me the same courtesy in return! My action did not lead IBM management to see the foolishness of the employment policy, but happily, others did.

IDENTIFYING THE PROBLEM BEHIND THE PROBLEM

As we all know, problems can be tricky. Just when we think we have figured out how to solve them, and we apply our solution, we realize we have, in fact, another problem. Let me explain.

Years ago, I did some basic research with my son while we piloted our plane at 20,000 feet. We measured the ability of volunteer test subjects to complete mathematical problems while using either carbon dioxide (CO_2), oxygen, or pressurized air. The results showed that these substances are equally effective in maintaining physical and mental function. However, a very much smaller volume of CO_2 than of oxygen is required to maintain the same level of function, because CO_2 itself regulates oxygen levels and, in effect, can fool the body into maintaining a higher oxygen level even when atmospheric oxygen is low. (This discovery led to the use of CO_2, in the form of dry ice packets, by skiers, mountain climbers, and others needing to accustom themselves to a higher altitude. And indirectly, it has led to a somewhat cleaner Mount Everest. Climbers no longer lug heavy oxygen canisters up the mountain, discarding the empties along the way.)

Sometime later, my wife and I flew my twin-engine aircraft to La Paz, Bolivia, which is 13,000 feet above sea level. The thin oxygen-poor air at that elevation caused me to become very ill. The only thing that gave me any relief was oxygen, but the only extra oxygen available to us in La Paz was the 20-minute emergency bottle in our plane, which was soon used up. The question was: Then what?

Initially, I thought, if I breathe twice as rapidly, I'll take in twice the oxygen. I had forgotten one thing: I'd also be exhaling twice as fast, and since we exhale carbon dioxide, I'd be losing twice as much CO_2 as normal. Carbon dioxide *is* a waste product, but the body needs it to regulate breathing and blood oxygen levels, so losing too much carbon dioxide by rapid breathing makes you just about as sick as not getting enough oxygen. In nonscientific parlance, I was damned if I did and damned if I didn't.

My initial plan wouldn't work unless I could also get more carbon dioxide. Where could I get more carbon dioxide in La Paz, Bolivia? At the local ice cream factory, of course. Dry ice is used to make ice cream, and dry ice is simply frozen carbon dioxide. At the factory, I asked them to fill my empty 20-minute emergency oxygen bottle with high-pressure liquefied dry ice, which they called Carbonico. That gave me enough CO_2 to support my extra breathing for the remainder of my stay in La Paz. My altitude sickness was cured.

My problem at La Paz turned out to be not so much the obvious one (I needed more oxygen) as a less obvious one (I needed more carbon dioxide). So the path to a solution was indirect rather than direct. I couldn't find an answer until I focused not on what I needed to get more of, but on what I needed to lose less of. To solve a problem with inventorship, you must always make sure you're looking at the right thing, and sometimes you must look at it in another way.

Similarly, I found the problem-behind-the-problem in another arena. As an employer, I have confronted the problem of unwarranted absenteeism. The obvious solution is to penalize employees who don't come to work when they should. But that doesn't solve the real problem: lack of incentive.

It took a snowstorm to lead me to a solution. On the morning following the storm (a minor one, resulting in only a couple of inches of the white stuff on the ground), only two people showed up at my company, the Safe Flight Instrument Corporation. The rest treated it as a "snow day" and stayed home. Rather than issuing a memorandum imposing a severe new penalty for missing work, I did the opposite. I instructed the vice president to announce that the two employees who had reported to work would be given double pay for that day. The very next week, we had a blizzard; eight inches of snow fell. Somehow every employee of the Safe Flight Instrument Corporation made it in to work that day. (See Figure 4.9.)

Figure 4.9   A simple incentive is sometimes all it takes to solve a problem.

Being Ahead of Your Time

Being at least one part visionary is a mark of all inventors, so never be afraid to be a little ahead of your time, even if it means a solution you discover is not adopted immediately, whether for reasons of cost, lack of awareness, or your "audience's" unwillingness to change. Eventually, technology usually catches up and makes the impractical suddenly practical. Inventorship should always be undertaken for the joy of the discovery and the satisfaction in the solution.

I was awarded a patent for one of my inventions, a new type of ski binding that, as good as the solution was, proved too expensive to market and manufacture—so far, that is. Nevertheless, I still take pride in this innovation and the way I developed it.

Those of you who are skiers will know that, in skiing accidents, most broken legs occur because the leg bends the wrong way or too far. Let's face it; our legs carry weight better than they withstand twisting or bending.

Ironically, I can best explain how legs get broken by using the term "arm," in the sense that it represents the fact that force is applied away from the pivot by a distance, that is, the "arm." For example, I can lift a small package with my arm outstretched from shoulder height; but, to lift a heavy suitcase, I probably will have to pick it up without extending my arm. With my arm outstretched, the force has more leverage on my muscles, on me. Similarly, a ski has a long arm, as compared to the small arm of a foot or the size of an ankle, so it has a lot of leverage. And when you have a lot of leverage, you tend to get breaks. Moreover, it's a fair distance from the toe to the heel of a ski boot. It's longer than the foot itself, because the boot sticks out both fore and aft. The longer the "arm," the greater the leverage, and hence the greater the chance of a break. (The Lang company made the longest boots I ever saw. Skiers who wore these boots so frequently broke their legs that these accidents came to be called "Lang bangs." Eventually, the boots were redesigned.)

Any fall you take while skiing can be bad, but a weighted fall exacerbates the problem. In an unweighted fall, you have the sensation of almost springing up from the snow. For instance, you might be at the top of a mogul, snow hillock, or snow bump. When the skis go up over the crest of a mogul, the force of your body weight on the ski is light, and there's little pressure on the skis. (It's just like going over a bump at high speed in a car. Your body tends to rise up from your seat.) You feel light because there's low force or weight on your skis. In this circumstance, you're unlikely to get a Lang bang. But, in the valleys between the moguls, where you "bottom out" and start to rise, the force is more than your body weight. The skis tend to dig down because of the combination of your weight and the force pushing down on them. If the skis twist, or you fall, with that pressure on them, the likelihood is that the solid connection between the skis and your boots will transmit great force to the bones of your leg or ankle. That force is transmitted with an "arm"—the arm of the ski going to your boot—and the length of the ski gives it a lot of leverage on your leg. If, at the same time, your ski isn't lined up precisely forward, the ski will have a "dug-in" twist to it, increasing the leveraging and twisting effects described.

In skiing, the forces that tend to hurt you have a big advantage: They've got leverage. Think about it: To break a twig, you don't pull or push it; you bend it. In the same way, pushing or pulling forces almost never break legs, whereas bending forces routinely do. As I've explained, that bending comes from forces applied over a spatial separation or from a distance, by an "arm." That prompted me to question: What doesn't have an arm? The answer: a point. Thus, my solution was to pivot the connection between the boot and the ski, so that the bending force couldn't be transmitted.

In the platform binding I designed, the skier stands on a ball under the center of his or her foot; and the shoe has a release at the toe and at the heel. When the twisting action exceeds a set limit, either the heel or the toe is released, precluding a "lock" on the skier's foot. Even if the skier were carrying a heavy weight on his or her back, or landing from a jump where the load was great, that would deliver just compression, or push, not the twist that breaks legs.

To test my theory, I asked an expert skier to do all sorts of fancy ski tricks and hotdogging stunts on these "pivot" bindings. At the end of his run, after racing down some challenging slopes, he deliberately crossed his skis and fell forward on his face. In ordinary bindings, this maneuver would have resulted in two broken ankles. Instead, he got up and walked away, leaving the two released skis behind, crossed in the snow.

Applying the Right Solution to the Problem

This chapter has described learning to define problems accurately so that we may become capable of applying the best solution to them. Fortunately, many problems bear the seeds of their own solution; we just have to learn to evaluate situations in inventive, innovative ways, so that we don't overlook or discount the most productive or valuable answer. Keep in mind the following inventorship points as we approach Chapter 5, where we look for sources of solutions:

- The existence of a solution does not imply that there is no longer a problem to be solved.

- When something isn't working, and you can't figure out why, try turning it on its side—or upside down, inside out, or backward—and see what happens.

- If you can't come up with a way to get from A to B, maybe you can get from A to C, then easily move back from C to B.

- Inventorship may mean inventing a new use for a solution to another problem.

5

The Source of Inventive Solutions

Invention breeds invention.
—RALPH WALDO EMERSON (1803–1882)

A ndy Rooney, the resident curmudgeonly verbal essayist on *60 Minutes,* the long-running CBS news magazine, launches many of his eclectic observations with a captivatingly simple phrase: "Ever notice . . . ?" It's a good question to apply to the process of inventorship. What do *you* notice? And what do you do with what you notice? Innovation often starts with just noticing things—those that work right and, just as importantly, those that "work wrong." As you may already have "noticed," in my own experience as an inventor, just observing, then investigating, has led to some rather remarkable ideas.

By now, I hope you're comfortable with evaluating situations and problems from an inventorship mind-set. The next part of the process involves learning how to go looking for inventive ideas and solutions. In this chapter, I will share with you some techniques that have worked pretty well for me over the years.

When you learn how to look at situations or objects through the eyes of an inventor, you will begin to examine such prosaic—and disparate—entities as water, to come up with a new kind of eyeglasses; welfare, to devise a revolutionary plan for tax reform; and worms, to imagine a design for a new supersonic aircraft. Sound bizarre? I'll explain each of these later in the chapter. Here, I want to elaborate on a concept touched on earlier: the importance of breaking down a complex issue or situation to find the most inventive approach to handling it.

Simplifying the Complex

Typically, even the most complex problems comprise a series of simpler problems. If we can break down complex problems into their components, they are less likely to break *us* down. To describe this process, I will narrate my experience in learning to defeat the phenomenon known as wind shear.

In 1975, a catastrophic crash at Kennedy Airport in New York introduced the public to the deadly force of wind shear. Soon after that disaster, I undertook an investigation of wind shear, which eventually led to the second most well-known of my inventions, Wind Shear Warning—now mandatory equipment on every airliner in the world. The most significant realization during that process was that wind shear was not one problem, but two; in the vernacular, wind shear exerts a "double whammy."

The first whammy is a change in the direction of the wind; instead of coming at the front of the plane, it hits it from above. The speed of the plane then drops, and the wings lose a portion of their lift. (In case you've forgotten your high school physics, the speed of the craft as it moves through the air generates the lift to keep it aloft.)

The second wind shear whammy happens at the same time as the first. It's called a downdraft. The cause is a weather phenomenon known as *virga*. A very high altitude rain cloud weights the air with water so that it first becomes a descending column of rain, and then, as the water evaporates at lower altitudes, turns into a downspout of air—the downdraft. Think of it as a "waterfall" of air rather than water. A waterfall pushes down anything in the water; a downdraft does the same to an airplane. When a plane has enough lift, it can resist the downdraft and keep flying, just as salmon with enough strength can swim up a waterfall. But in wind shear, because of the first whammy (the change in the direction of the wind), the plane no longer has the lift it needs to do that. When a plane that is too close to the ground comes in contact with wind shear, there is a very good chance that the plane will go down.

If either of these wind conditions occurs alone, it is usually survivable, but together they generate a particularly dangerous situation. Wind shear is a force of nature that cannot be prevented from occurring. Thus, the only way to defeat it would be to know it is coming, increase airspeed, and start climbing before it hits.

I set out to invent a device that would give pilots the advance warning they needed to avert disaster. Two factors made the invention possible for me: (1) I recognized that wind shear was a force of two components; and (2) I figured out how to measure each of those components when they start to happen.

I knew that losing power or lift while flying doesn't inevitably mean a crash. Today's multiengined aircraft are designed to stay aloft when one engine loses power. Lose any *more* power, however, and there's trouble. Therefore, I set out to design a device that would measure the results of two separate losses (altitude and lift), and then add them together. If the total loss exceeded the power of a single engine, the device would set off an alarm. The alarm would tell the pilot to use emergency power, enabling the plane to escape an impending crash. (See Figure 5.1.)

I learned a valuable lesson from my investigation into wind shear, and the subsequent work leading to my invention of

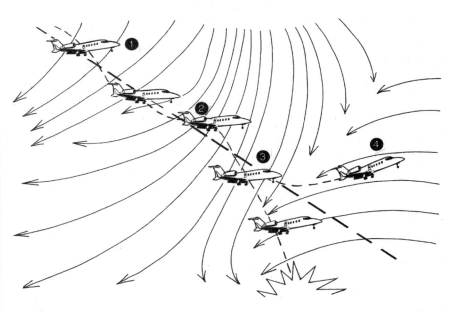

Figure 5.1 Wind shear. (1) The aircraft experiences a change in wind direction—first head-on, then from above. (2) The pilot compensates, and the aircraft flies above its intended flight path. (3) The aircraft experiences a downdraft, forcing it well below its intended flight path, and—without the advance Wind Shear Warning—possibly into the ground. (4) With the advance Wind Shear Warning, the pilot adds emergency power and escapes a dangerous situation.

USING THE SIMPLE TO DO MORE

I've never been a proponent of the cliché: "Necessity is the mother of invention." In my experience, *innovation* earns that title; however, I will concede that necessity can be a nurturing foster parent.

In previous chapters, I've referred to a sailboat trip my wife and I took, from New York to Bermuda. As difficult as that experience was, it taught me a lot about the heart and soul of inventorship. Another lesson I learned on that stormy journey has relevance to this section, too: It taught me the value of "simple" innovation, of learning to do more with less.

One of the problems we encountered was that our sailboat's engine failed on our first day out. The stormy weather had not yet arrived. The sea was beautiful and quiet. Being on a sailboat, we still had motive power; but we had no way to recharge the boat's batteries, which powered our instruments, our radio, our lights, and the refrigerator. Naturally, because it was the first day of the trip, we considered turning back. But the radio made it clear that turning back was not an option; the storm was coming up quickly, on our tail. We were told, "Keep going forward as fast as you can." Fortunately, we had a good wind and could do that pretty well, at least for a while.

Our primary concern at this point was watching for the big ships. When you're a small boat in a big pond—in this case, the Atlantic Ocean—you must watch out for big bruisers that are sharing the same water. Large commercial vessels don't tend to look out for small craft or even pay much attention to their radio messages, so the burden is on the little boats to watch for the big ones and try to stay out of their way.

I had a battery-operated tape recorder with a timer. I recorded a message announcing the presence of our boat (and the fact that I was transmitting at very low wattage) and set the timer to replay the message every five minutes. I placed it by the boat's radio,

(continued)

Making do with less may be just the innovative step you need to take.

which I set to transmit at just one watt—about the amount of power used to operate a tiny "penlight" flashlight. It was all the electric power we could spare, under the circumstances. In my recorded message, I made special note of the very low wattage on which it was being broadcast.

My message got through to the big ships remarkably well, and some responded by giving us assurances of safe clearance. The announcement that I was using so little power was a clear indication to them that my boat must be close; otherwise, they wouldn't have been able to hear it. Because of this, the message was hard to ignore. With a little ingenuity, sometimes you can generate innovative solutions to difficult problems.

Wind Shear Warning: When a solution to a complex problem continues to be elusive, identify the various parts of the problem, and begin by attacking just one of those parts. Then go on to the next part, and then the next . . .

Changing Point of View

One of the most effective ways I've found to keep my "noticing capability" in tip-top shape is to take a second look, but from a different viewpoint. Here are three examples of how this practice stimulated my own innovative thoughts.

SPEAKING A COMMON LANGUAGE

Educators have known for some time now that children learn best, and more quickly, when they're having fun. The same holds true for adults. This simple observation led me to another invention. While visiting Sweden several years ago, I noticed that many of the Swedish cab drivers spoke excellent English. I complimented one of them on his mastery of the language, and asked whether he had studied it in school. "No," he said, proudly. "I learned English at the movies." It turned out that most of the pictures shown in Sweden were American-made, and they were not dubbed, so Clint Eastwood's famous challenge, "Make my day," sounded the same in a cinema in Stockholm as in a multiplex in Spokane.

I couldn't stop thinking about what my cab driver said; I knew he was on to something. If people, especially young people, need to learn another language, why not make the process fun?

The result of my serendipitous conversation with a Swedish hack driver was an invention I called SoundTitles. I took a mix of feature films, training films, and filmed TV shows, and added audible—not visual—subtitles. Thus, rather than a hard-to-read printed translation at the bottom of the screen, the films had a sort of second soundtrack in the language being studied. (See Figure 5.2.)

I launched the invention using Spanish because in my part of the country (the Northeast) the immigrant Hispanic population is very large and there's an obvious need to make learning English easier and more fun. The narrative was less a translation of each piece of dialogue than a brief description of what was going on in the movie. It was like having a bilingual friend sitting in the next seat in the theater and summarizing the plot aloud. This description was inserted during intervals in the actual dialogue. I also patented a fade-in/fade-out process that made the SoundTitles as unobtrusive as possible. Humana Films later applied the idea to 10 training films for

Figure 5.2 Learning a foreign language by going to the movies is an innovation everyone can enjoy.

Hispanics. The films were very effective, but the technique was intended for pleasure as well as teaching. For instance, the feature-length film *La Illegal*, which won several prizes at the Cannes Film Festival, was about half English and half Spanish, a language mix required by the plot. The SoundTitles technique made this excellent film equally accessible to both Spanish- and English-speaking viewers.

"SEEING" WHAT THEY'RE SAYING

In Chapter 4, I described adapting an innovation to solve a different problem. That's precisely what I was able to do after considering my SoundTitles idea from another point of view. Essentially, I made speech visible *without* writing it down.

Specifically, my idea was to design eyeglasses that, in effect, could help people *hear* better. How? By (1) building a receiving unit into the temples of the eyeglass frames, (2) adding a frequency analyzer that could translate, into a visual display, the sounds picked up by the hearing aid, and (3) outfitting the bottom edge of the lenses with a light-emitting diode (LED) or liquid crystal display (LCD) screen. (See Figure 5.3.)

The screen would display lines corresponding to the sounds, as a sort of scrolling billboard or electronic ticker tape. This display would convey not only words, but also characteristics of the sound, such as pitch. For example, a user would be able to tell a female voice from a male voice and even be able to identify the speaker. People wearing the glasses would, of course, first have to learn to read and interpret the displayed

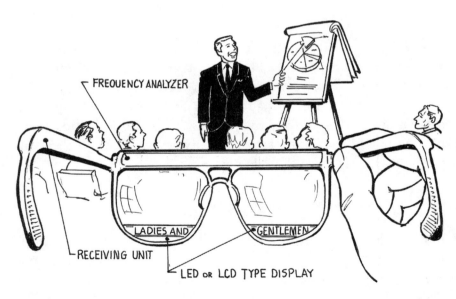

Figure 5.3 Making speech visible is an innovative way to assist those who are hearing-impaired.

CHANGING THE VANTAGE POINT

As I write this, it is autumn in the Northeast and the leaves are beginning to turn. Here, people waste no time heading out into the countryside to admire the fall leaves in their flamboyant display of color, especially during the last weeks before winter sets in. I've taken the same journey many times, but it wasn't until last year that I believe I really *saw* the autumn colors.

I was flying to my country home in Connecticut in a helicopter piloted by my son. We were only about 200 feet above the trees, the colors had just reached their peak, and I was astonished at what I saw. The colors were of an intensity and variation I had never before experienced. I wondered whether there was something special about that particular season. Had there been more—or less—rain, or more warm—or cool—summer days than usual? After thinking about it a while, I realized it had nothing to do with the state of the environment. What was special was my vantage point. Like most people, I had always seen the leaves from the ground, usually from a road cut through the woods.

When you're looking *up* at the leaves from ground level, you see their undersides, not their faces. And from a road, you're seeing only those species that can adapt to survival at the edge of a cut, which means you're seeing the leaves of fewer kinds of trees.

In contrast, when you're above the trees, looking down, you see them full-face in all their glory, without shadows. The view I had reminded me of a French Pointillist painting. Every leaf was a distinct dot of color.

If you're wondering what this has to do with inventorship, it's this: finding another way may require another point of view. If there's a problem for which you can't *see* a solution, maybe the obstacle is your vantage point. As long as you look at a problem in the same old way, there's only so much you'll be able to see. Approach the problem from a completely different angle, and you may easily find the "other" way.

(continued)

CHANGING THE VANTAGE POINT *(CONTINUED)*

Changing your point of view may lead to new insights. For instance, the colors of autumn leaves appear more intense and brilliant when seen from the unusual high vantage point of a helicopter than from the accustomed ground-level.

lines, a process similar to learning shorthand. A deaf person could then converse without lip reading or signing. After a friend had finished talking, he or she could respond, "I see your point," and mean it *literally*. With this invention, the deaf could see not just words, but also music. After all, words are only one manifestation of sound, and the glasses would allow users to "see" *any* sound. Think about it: talented musicians can look at a score on paper and "hear" the music in their heads, even if they've never played or heard the composition before, because what the ears hear and what the eyes see transmit meaning to the brain.

If the eye can "hear" music in the form of painting (as I explained in an earlier chapter) and the ear can "see" color in the form of music, we have a transposition of senses. Eventually, I predict, we will achieve hearing aids that operate through sight, and seeing aids that operate through sound.

Bending the View

The third example of how noticing things from another point of view can lead to innovative thoughts occurred to me while swimming under water. Because water bends the light that we see, our vision is distorted under water. Corrective lenses do the same thing, but they improve our vision rather than distort it.

I wondered whether it would be possible to take water's capacity to bend light and use it to improve our vision under water. Would we then see farther underwater than the usual hand-before-one's-face? If corrective lenses can bend light to make it easier to see, why couldn't water do the same thing?

I designed some plastic glasses with curved lenses (plastic does not bend light). Because the surface of our eyeballs is convex, and the plastic lenses make the water against them concave, these lenses, when worn under water, alter the shape of the water enough to bend the light in a way that corrects the distorted vision. (See Figure 5.4.) (You may be wondering how the shape of the water could be changed, since we commonly think of water as something amorphous. But in fact, a shape is a boundary. If you put your hand into the water, the water "bounding" or surrounding your hand has a new shape: the shape of your hand. We don't think of air as having a shape either, but we've already talked about the way airplane wings shape the air as they move through it.) Visibility under water was greatly improved with these plastic lenses, and my kids and their friends had a lot of fun with them. They were not produced commercially, but might make a good prize for inclusion with Cracker Jack.

Figure 5.4 Noticing one phenomenon may lead to explaining or clarifying another. Using the capacity of water to bend light *improved* our vision underwater instead of distorting it.

"REFLECTING" ON THE PROBLEM

Try looking at a problem from the inside out, or look at its reflection. This productive approach was useful in solving opposite problems at two different skating rinks.

About 30 years ago, the White Plains Department of Recreation built a beautiful rink that should have brought droves of people out to enjoy the sport, but it had a baffling drawback. Despite ample refrigeration capability, which solidly froze most of the skating surface, the ice near the edges stayed wet and spongy. Beginners, who had to skate near the handrail so they could hang on from time to time, found that the mushy edge made their efforts even more difficult and discouraging.

A friend who was involved in the ice rink told me of the problem and I "reflected" on it. That brought to mind something else that reflected on it: the sun.

The handrail was supported by sheet aluminum, which had the same effect on the ice as the reflectors people sometimes use (perhaps foolishly) to get a quick sunburn on their faces. (Although the rink was much cooler than the tanning environment would be, the increment of the rise in temperature is about the same in either case.)

I suggested that if the sheet-aluminum rail support were painted a dark color, to stop it from reflecting, the ice near it would be just as firm as the rest. They did that, and since then, Mama, Papa, the kiddies, and all the other novice skaters can comfortably skate near the rail, without having their skates sink into soft ice.

At the other skating rink, the problem was the opposite: not reflected heat, but absorbed heat, and the *cure* was reflection.

When my children were young, we lived too far from White Plains for convenient skating at the new rink there, but they still wanted to skate, so we decided to make our own rink. I had learned that, in Canada, "home-made" ice rinks are commonly

(continued)

"REFLECTING" ON THE PROBLEM *(CONTINUED)*

Inventorship can consist in stopping the effect of a force by absorbing it. In this case the warming capacity of the sun's rays was "turned off" by painting the reflective surface a dark color.

made by spraying water from a garden hose upward so that it arcs out into the frigid air, where it freezes into tiny ice droplets. When these droplets land on a porous surface, like a tennis court, a lawn, or even bare dirt, they don't run through the surface or make it muddy. Instead, they immediately fuse, forming beautiful, solid ice. It's not white ice, because there's no air in it. It's what we call "black ice"—clear and transparent, not translucent. It's the best ice for skating, as it's extremely durable.

We tried this on our tennis court, and soon had a beautiful rink. The one problem, we discovered, was that, on sunny days, the rink melted and we had to start all over again. Even if the air temperature is below freezing, the warmth of the sun will melt this transparent ice. The sun's rays go right through it and are absorbed in the subsurface, which then melts the ice from the bottom up. (This is called the "greenhouse effect," and it works exactly the same way with glass as with ice. The sun coming through a window doesn't warm the windowpane; it warms you, if you happen to be sitting inside near it.)

As we saw with the White Plains rink and the problem of reflected heat, wherever the sun's rays stop, that's where the warmth is. Using the same "reflective" thought process, I concluded that our problem would be solved if we didn't let the rays stop on the surface of the court under the ice, where they were transformed into heat, but instead, laid aluminum foil on the bare court to reflect them back through the ice into the air. We covered the court with Reynolds Wrap. Then we did our spraying (usually, at night, because it's colder). The next day, the sun blazed down on our rink, but instead of melting the ice, its rays warmed the bottoms of the skaters.

The aluminum-foil treatment was so effective that even a temperature a few degrees above freezing still made, and held, ice. That Reynolds Wrap gave us many a day of great skating.

(continued)

Inventorship can mean *reflecting* on a problem in order to *deflect* the cause. With our aluminum-foil treatment, we deflected the sun's rays back into the atmosphere before their heat could be absorbed in the subsurface and melt our skating-rink ice.

Lowering the Tech

Since the early 1990s, most inventions we have heard or read about have seemed to be dependent on computers and related technologies. While not discounting the value or importance of these developments, I fear they have impacted the natural tendency of both children and adults to think and act inventively. The purpose of this section is to serve as a reminder that, to be innovative, advanced training in computer programming is not required. Many of the best solutions—even today—are simple and low-tech.

PREVENTING ACCIDENTS

Radio is now considered old (that is, low) technology. Yet it is important to remember that radio was once heralded as a marvel, not unlike the way people react to the technologies of today. In fact, many modern technological developments would not have been possible without the understanding gained from working with radio waves. This now-humble technology can still hold its own, as this inventive implementation demonstrates.

Hitting power lines has always been a major cause of helicopter accidents. This seemed a solvable problem to me. What power lines needed, I thought, was an advance warning mechanism to alert chopper pilots when they were getting too close. I found the solution by listening to the radio.

You've probably noticed that when you are listening to AM radio in your car, and you drive near power lines, you hear static. I turned that annoyance into a valuable warning system. I installed an ordinary AM radio receiver in a helicopter, together with a device that would convert static into audible clicking, like that of a Geiger counter. When a helicopter comes within static range of a power line, the pilot hears the warning in time to turn away, thus preventing a crash. (See Figure 5.5.) The Power Line Avoidance System is a commercially successful new technology that is very much in demand.

Landing in fog is another dangerous problem for helicopter pilots. If, during periods of poor visibility, they allow their craft to drift even slightly, they'll miss their landing pad. Drift is caused by wind currents, and, given the high speeds of the helicopter rotors (the spinning blades that give the helicopter its lift), the effect of the wind current increases geometrically. That means the pilot must know his or her speed exactly.

Figure 5.5 The static that power lines generate on the AM radio can be used to warn helicopter pilots of the danger and help them avoid accidental contact.

My solution again involved a new use for a piece of older technology—in this case, the simple wind detector, used for years in navigation, weather reporting, and so on. I put a wind detector on the tip of the helicopter rotor blade, where it measures differences in air speed as small as one mile per hour between the upwind and downwind portion of the blade's rotation. That increment is hard to register any other way, but it can make a big difference in measuring wind current. Using this Drift Detection System in combination with other navigation tools can ensure that the helicopter is always exactly where the pilot intends it to be.

PINNING DOWN THE IMPORTANT THINGS

This story should convince you that no technology is so "low" that an inventive mind cannot make use of it. I learned this lesson from an ordinary clothespin.

Antennas on airplanes today are short, but years ago as much as 200 feet of antenna wire were strung out behind aircraft. Pilots had to pull the wire into the plane with a reel—that is, if they remembered it was out there, which was not always easy to do when landing and having lots of other things to pay attention to. Several times, I forgot about the antenna and it tore off during landing, forcing me to buy a new one. My solution couldn't have been more low-tech—or imaginative: I simply clipped a clothespin on to the handle of the antenna reel. When I unreeled the antenna, I would move the clothespin to the throttles. (When you land a plane, you always pull the throttle back, and when I did that, the clothespin was there.) That little wooden gizmo served as a reminder that I had to crank in the antenna.

I later called clothespins into service on the water as well. When you bring a sailboat with a centerboard into harbor, you have to remember to pull the centerboard up before you hit shallow water. Well, if I couldn't remember to pull in my 200-foot antenna, I surely couldn't remember my five-foot centerboard. But I did remember my clothespin. I put a clothespin on the engine throttle. When I'd start the engine to maneuver the craft to the dock, and I turned to the throttle, I'd see the clothespin and remember to take care of the centerboard.

DIGGING DEEP FOR IDEAS

Think you can't get much lower than a clothespin, technologically speaking? Well, I can, and I did. The source of one of my patented concepts was the lowly earthworm. It struck me as remarkable that this "primitive" creature is able to move through soil it doesn't have the strength to push aside. The earthworm accomplishes this by swallowing the dirt and then excreting it at the other end. I reasoned that what works for the worm below the ground might work high above the ground as well.

When pilots fly above the speed of sound, they run into a problem. The aircraft can't push the air in front of it out of the way fast enough. That creates enormous resistance to the forward motion of the plane—as well as a loud thunderclap better known as a *sonic boom*. But just suppose, mimicking the earthworm, it was possible to open a "mouth," or hole, in the front of the plane, and extend it like a tunnel all the way to the rear. That would allow the craft to "swallow" some of the air instead of having to push it aside. And if it were possible to curve the inner walls of the tunnel so it was narrower in the middle of the airplane than at either end, the entering air would be compressed and be able to move through easily and exit at the tail. This would eliminate the sonic boom. (See Figure 5.6.)

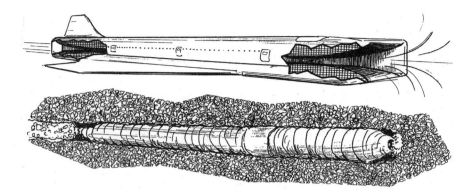

Figure 5.6 The earthworm inspired an innovative design for a supersonic aircraft.

I built a prototype based on these principles, and had it tested at two U.S. military proving grounds. It was found to have less aerodynamic drag than anything either the Air Force or the Navy had ever tested. An artist's conception of the application of this idea hangs on the wall of my office. It's a 250-passenger supersonic airliner that would fly 50 percent faster than the Concorde, but consume less fuel per passenger than a Boeing 747. It could also fly safely and quietly over land because it would avoid creating the usual shock wave. This supersonic aircraft could get travelers from New York's Kennedy Airport to Tokyo in three and a half hours.

Boeing bought my concept in 1996, but, unfortunately, it is still considered too expensive to develop, so it may be a while before you can book a seat on a "flying earthworm." That doesn't change the fact that a quite revolutionary idea came to me from contemplating an extremely simple, earthbound creature. I think you see the point of this section: With a little imagination, some of the least complex things can become the most useful.

Paying Attention

Many dramatic innovations are really applications of very simple ideas—proof that you don't have to be in a corporate research laboratory to practice inventorship. Your laboratory is the world around you.

Though the concept of atomic accelerators, which create energy by causing atomic particles moving in opposite directions to collide with each other, is considered a modern phenomenon, I recognized it much earlier, while playing with an electric train set when I was seven years old.

I had arranged my train track in a figure eight. I had two trains, and by turning the power all the way up, they moved at what I judged to be about five miles per hour (5 mph). As is typical of young boys, one of my favorite strategies was to point the trains in opposite directions and have them crash. By doing this, I learned that two trains coming at each other at 5 mph collide with their combined force. In other words, I was able to get, out of the crash of two 5-mph trains, a 10-mph effect, which was nominally beyond the capabilities of my equipment. And this was accomplished just by running the trains another way. (See Figure 5.7.)

Years later, I applied the lesson of the toy trains to nuclear physics. I attended, at Princeton University, a lecture on fusion energy. The top particle speeds achievable in Princeton's cyclotron were discussed as a limiting factor in the ongoing research. Afterward, I spoke to the lecturing professor. Suppose a cyclotron took the form of a figure eight and propelled two particles traveling in opposite directions. Would not the force

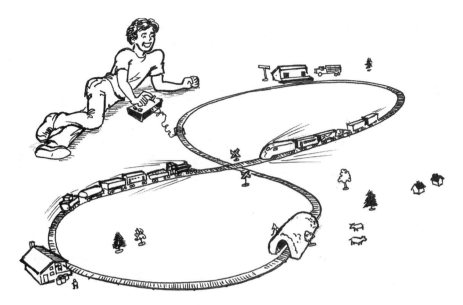

Figure 5.7   What do toy trains and nuclear physics have in common? More than you might think.

of that collision be double that of an individual particle's speed? I asked him. These principles, later used in a modified cyclotron, moved fusion energy research to a new level.

As I explained in Chapter 2 ("The Age of Innovation"), childhood games and toys stimulate the imagination for a lifetime. Many inventors have been inspired by toys. Elmer Ambrose Sperry invented a device that would enable airplanes to "fly blind," and the idea for it came from a toy gyroscope.

In the early years of aviation, when planes ascended into the clouds, pilots would often lose control and crash. The reason was made clear when aviation pioneer Wolfgang Langewiesche blindfolded a pigeon and then released it high up in the air. Without its sight, the pigeon could no longer fly.

We now know that the physical sense of motion is not enough to guide us when we fly. If, for example, you're in a jet with the shades drawn, watching a movie or reading a book, you won't be able to sense which way the plane is turning. You have to be able to see something external as a reference point. When a plane is turning and the pilot doesn't know it, he or she can lose control. At one time, this seemed like an unsolvable problem, and the future of aviation depended on finding the solution.

Gyroscopes have "spin" inertia that resists tipping. (Recall how a spinning top will right itself if given a light push to one side.) Sperry concluded that by resisting the gyroscope's turning with a spring (the amount of spring compression would correspond to the rate of turn), and then connecting a pointer to the spring, the pointer could indicate that a plane was turning even when there was no visual evidence of the turn. That would cue the pilot to take corrective action to keep the craft under control. Sperry's gyroscopic turn indicator allowed pilots to fly safely without having to see either the ground or the horizon. His toy gyroscope became the foundation of a giant industry.

Reconsidering the Obvious

Another of my insights into problem solving through inventorship first came to me as a child, not through play with a toy, but through childhood reading—in this case, *The Book of Knowledge*, a wonderful children's encyclopedia first published in 1910. The tale I refer to told of a thirsty bird on the shore of an island. The only fresh water it could find was rainwater that had welled up inside a deep shell. Perching on the edge of the shell, the bird tried and tried again to reach the water, to no avail.

The bird couldn't lower itself, but it could raise the water. It flew off and came back with a pebble, which it dropped into the shell. It did the same thing many times. The pebbles "displaced" the water, raising its level high enough to allow the bird to drink. (See Figure 5.8.)

Remember Archimedes? This bird and Archimedes had some things in common—observations about the displacement of water, and inventiveness!

MAKING BOOK ON IT

Much later, I applied the bird's lesson when, out of frustration, I began to reconsider something obvious. The frustration occurred when I attempted to read a 1,000-page book while lying on my side in bed, which is how many of us prefer to read. I first tried propping the heavy volume up against my knees. That didn't work; frankly, nothing else did either. It didn't take long before I started thinking, "There's got to be another way

Figure 5.8 Reconsidering the obvious—even innovative birds do it!

to do this. Why should there be only one obvious way to produce a book and to number pages?"

If we just accept the "obvious," we're forgetting the principle of another way. Why *shouldn't* there be more than one way
to number pages? Why *can't* there be more than one way to
decide what goes on the next page and what goes somewhere
else? Isn't it possible to create, if not a better book, at least an
alternative book for special circumstances? So I set about to
prove there was an equally viable, alternative way of producing
one of the oldest forms of technology: the printed book.

I am most comfortable lying on my left side when I read in
bed. That works fine when I'm reading the right-hand, odd-
numbered pages; the pages I've already read can lie flat against
the bed. But trying to read a left-hand, even-numbered page in
this position can be frustrating. An alternative would be to turn
over every time a page is turned, but that's hardly relaxing, and
would drive the loved one sharing the bed crazy.

Why not, as an option, I reasoned, print pages 1 through
500 on all the right-hand pages, then print pages 501 through
1,000 upside-down on the reverse sides of these pages. (See
Figure 5.9.) As to the marketability of such an innovation, I
suggest that it might be sold to those who are hospitalized or
bedridden. This innovative format might also be easier for all
readers, once they became comfortable with it. We can change
our conventions and habits if we find another way that works
better. (How many of us would be willing to go back to type-
writers, now that we have computers, which make correcting
our work so much easier?)

Amplifying the possibilities for pagination of books, I submit
another idea: Imagine a book designed for a child and a parent
to read together—specifically, a book with one story, but alter-
nating pages for the parent and the child. The child's pages
might always be the recto (right-hand) pages, and the parent's

Figure 5.9 A one-sided book would be much easier to read in bed. The first half of the book would be printed on the right-hand pages, and the second half would be printed on the left-hand pages.

pages the verso (left-hand) pages following the child's pages. (See Figure 5.10.)

The child's pages would tell the story in the traditional manner—that is, using very basic vocabulary and graphics appropriate to the age group of the child. The parent's pages, written at an adult's reading level, would amplify the story, making it more interesting for the child. Key words in the adult-level story would constitute the entire text of the child's story, and these words would be read together. New vocabulary would thus be introduced and its learning would be reinforced. Ideally, these books would be designed as a series, with each

Figure 5.10 Another innovation for books would be to print a parent's page opposite a child's page, to make reading together a richer experience.

new book building from the vocabulary the child learned in reading the previous one.

With this type of book, the child and the parent could get more out of the shared reading experience. Each would participate at his or her own level. Reading such a book would also bring parent and child together—literally as well as figuratively. When one reads from the left-hand page and the other from the right-hand page of the same book, closeness is ensured. The experience would be both pleasurable and productive.

And lest you think such innovations are cost-prohibitive, computerization is making it possible to investigate alternative cost-effective book-production processes. This is another example of how one innovation or solution can prompt another in various professions or fields of study.

Piecing Together the Solution

In this section, I expand on the various ways to approach problems inventively. I've talked in earlier chapters about doing a "two-step"—going sideward or backward in order to move forward—and doing the opposite of the obvious. Here, I want to discuss inventive solutions that don't come in one piece, nicely assembled, and ready to use out of the box. Sometimes they're more like a kit—without any instructions. I can best illustrate this form of inventorship by telling you about the time I got off an island by using a lawnmower.

If you've ever forgotten to turn off your headlights and have come back to a car that wouldn't start because of a dead battery, you'll appreciate this story. It's the same problem, but on a slightly larger scale. In this case, when my wife and I flew a small jet to Martha's Vineyard for a week's vacation and wound up with no battery power, calling Triple-A was out of the question. Luckily, inventorship was not.

As in a car, a jet's engine is started electrically, from a battery. But jets use nickel cadmium (NiCad) batteries, which can produce tremendous power output. Starting a jet engine requires about 1,000 amperes of power. (By way of comparison, your home probably has a total of about 100 amps of power. If you happened to find a jet with a dead battery in your backyard one day, you'd have to wire up your own and nine of your neighbors' homes to have enough power to move it out of there.) Despite the formidable capabilities of the NiCad batteries, jet engines require so much power that it is possible to start the engine only once before having to recharge the battery.

We had landed our plane at an unattended airstrip on the island, and thought nothing more about it until it was time to go home. But, as we learned, a nice little convenience provided by the aircraft manufacturer would cause us a whole lot of trouble.

At the top of the steps leading into the plane was a courtesy entrance light designed to remain on even after the engine had been shut off. Many cars have a similar feature, but these lights turn off automatically after a few seconds. In a plane, the pilot must extinguish the light manually. Well, it was daylight when we left the place, so neither of us noticed the light. Hence, upon our return at the end of the week, we had a jet with a dead battery.

As I mentioned, this little airstrip was deserted—no truck, no other planes, no people, no nothing. Just my wife and I, the two cars we had been driving on the island, and an empty hangar. We needed a minimum of 28 volts of battery power to start the jet's engine, and each car's electrical system and battery yielded only 12 volts. We needed four additional volts.

I forgot to mention: there was one other thing at the airstrip—a lawnmower, and it would prove to be instrumental in getting us off the island. Here's the solution I pieced together. I found a roll of electrical wire in the hangar, the kind that's used on doorbells in homes. I used it to connect the batteries of the two cars. That gave us a total of 24 volts. Then I linked up the little lawnmower, which had a six-volt system and battery. Total: 30 volts, two more than I needed. (See Figure 5.11.)

Upon starting up the lawnmower and both cars, we got enough juice to light the jet's headlights—but not to start the engines. So we left the engines of the three land vehicles running for an hour and a half. By that time, they had charged the jet's battery to a point at which I could start one of the jet's two engines, which I then let run until it had charged the battery enough to start the second.

This was a prime example of a solution cobbled together out of whatever was handy, and making it work another way.

Figure 5.11 Solving some problems may require a combination of several different factors.

ADDING SOMETHING MORE

Often, an innovative combination requires the addition of a new element or of more of the same innovation. I proved that lesson with my Stall Warning device, described earlier.

Landing a plane on an aircraft carrier is one of the trickiest maneuvers a pilot can be called upon to do. Though the carriers are huge, to a pilot descending from the sky to make a landing, they can seem but a speck in the ocean. To land successfully on a carrier deck, the plane must be at just the right attitude and speed; there's very little margin for error. The pilot must be precisely the correct percentage above the "stall speed" in order to stop the plane *on the deck*—rather than *in the water.*

It turned out that what made this action less treacherous was not one, but five of my Stall Warning indicators, each set apart from the next by one percent of the "landing" speed, and the middle one set for exactly the correct figure. (See

Figure 5.12.) The principle is like the rings around the bull's-eye on a target, or the warning signs before the full stop at a tollbooth. Such a graduated approach gives a measure of closeness to the center, enabling a more precise "aim" at the goal. We ran a landing test with the five-indicator setup, and it worked perfectly.

I solved another aircraft carrier problem with the addition of a bump. In World War II, we had to launch fighter planes

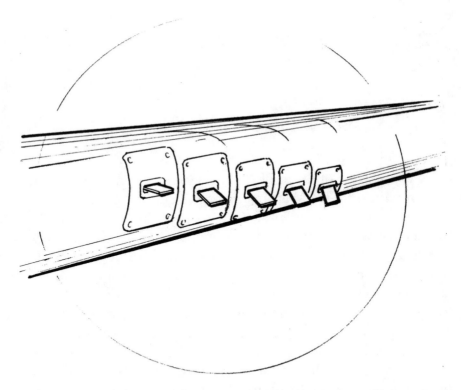

Figure 5.12 Installing several identical Stall Warning devices on the wing—set above and below, as well as precisely *at* the target speed for landing on an aircraft carrier—provided a great advantage.

from very small aircraft carriers, and the short runways created a problem. A plane couldn't always achieve flying speed before reaching the end of the shortened runways; and when it didn't, it would plunge over the end, usually to be subsequently run over by the ship. My solution was to set in, at the end of the runway, what was essentially a speed bump, like those found in parking lots everywhere today. When a not-yet-airborne plane hit the bump, the plane would be deflected upward, giving it the extra boost that could mean a successful launch. (See Figure 5.13.) It worked. Now almost all small aircraft carriers have a bump at the end of the takeoff runway.

Figure 5.13   Adding a bump created the necessary boost for a safe takeoff.

I confess, however, I may not really have been the originator of the bump idea. Charles Lindbergh is said to have installed a log at the end of the runway he used for his famous trans-Atlantic flight's takeoff.

Solving Simply, Simply Solving

In his book *Innovation and Entrepreneurship* (New York: Harper-Business, 1993), management guru and best-selling author Peter Drucker wrote, "All effective innovations are breathtakingly simple. Indeed, the greatest praise an innovation can receive is for people to say: 'This is obvious. Why didn't I think of it?'"

I think Drucker's quote sums up this chapter effectively: Your invention, idea, or solution to a problem could easily become apparent to you in a taxicab, in an elevator, in the parking lot at Burger King, or in any other place—as long as you continue to pay attention and to *notice*.

As we move forward, add the following inventorship points to your memory bank:

- Innovation often starts with just noticing things—those that work right and, just as importantly, those that "work wrong."
- If you break down complex problems into their components, they are less likely to break you down.
- To notice effectively, always take a second look, but from a different viewpoint.
- Reconsider the obvious.

6

Mind over Machine

Imagination rules the world.
—Napoleon Bonaparte (1769–1821)

One much-hyped component of our so-called "Information Age" is the concept of "artificial intelligence." The relevant word in that term, from my point of view and for the purposes of the theme of this book, is "artificial." It is the appropriate word because although computers are remarkable in many ways, they are still machines, and machines will never be capable of practicing inventorship. Artificial intelligence is not human intelligence, which comprises imagination. Because we have imagination, we can figure machines out, but they cannot do the same to us.

The purpose of this chapter is to refocus attention on the human imagination as a primary component of the inventorship process. Understandably, people today are caught up in the computer revolution because it is having such an impact on all aspects of our lives. However, to develop the inventorship mind-set requires stepping back from the world of machines and into our own heads.

Leveling the Playing Field

It's easy to become overawed by the capabilities of computers. After all, we live in an age when computers can even beat humans at chess. But I'm here to tell you: that's less impressive than it sounds. In chess, you can move only in accordance with certain rules. And, given the number of spaces on the board and the number of pieces, a finite—albeit large—number of moves can be made at any time. Naturally, that kind of well-ordered universe is ideal for the capabilities of computers.

But, as we all know, our world is significantly more complex than a chessboard. Problem solving in our complex world is still something we can do a lot better than computers, because we humans can look for another way. Let me give you two examples.

CUTTING A DEAL

I began to recognize the limitations of "smart" machines even as a young boy, and this recognition has been important to my development as a professional inventor and as an inventive thinker.

When my brother and I were young, we had a mechanical card table; it dealt cards automatically. This "smart" machine dazzled my older brother. He had bought it to play bridge.

When he inserted a deck, the machine could deal trillions of combinations of 13-card hands (because there are 13 cards in a suit).

As a much younger brother, I was told, "Hands off." That didn't stop me from being curious. As you can probably guess by now, I was determined to figure out how it worked. Ignoring my brother's admonition not to touch the table, one day, before he and his friends came down to play, I lifted the top off the table, and saw that the "dealing" machine was, essentially, a cam, which governed where the cards were dealt. With each new deal, the cam slipped over one slot so it would not distribute the cards in the same way twice in a row.

I knew I was on to something: I opened a new deck (the cards were in the standard order), and put it into the machine. I made a mark on the cam at its initial position. I then noted which table pocket each card went into in the first deal. This gave me the "before" and "after" for this specific setting of the cam, which told me how any deal, if started with the cam at that initial position, was going to come out. I unscrewed the cam and reset it to that same position. Therefore, the next time the machine was used, it would make the same distribution of cards.

Next, I sorted all the decks in a certain sequence. In this sequence, and with the cam at that specific starting position, the machine would deal four hands as follows: 13 clubs, 13 diamonds, 13 hearts, and 13 spades. Then I replaced the top of the table.

With my mind-over-machine experiment set to run, I laid down on the couch and pretended to be asleep when my brother came in with his card-playing buddies. He was boasting about his marvelous machine that could deal trillions of *unpredictable* hands. He put a deck in, turned on the machine, and they sat down to play. My brother was promptly dealt a hand of

13 clubs. Each of his friends received a hand that was equally homogeneous. (See Figure 6.1.)

Dumbfounded and angry, my brother, who had spent $40 on his mechanical marvel, swore he would never use the machine again. To this day, I have never told him how it happened. He is now 91 years old and will learn the truth when he reads this book.

The moral is: Given identical circumstances, a machine will produce identical results every time. Our brains, however, may come up with a *different* response to the same circumstances each time they are present. That flexibility makes inventorship possible—and exciting.

Figure 6.1   I proved that a machine is no match for the inventive human mind.

COUNTING CARDS

I also came up with a simpler method of remembering what cards remained in the deck at any given moment, which is what professional gamblers teach themselves to do. My method was to group the cards according to their importance to winning; in other words, I emphasized the relative value of groups of cards. I classified those with a value of 10 and higher, including the ace, as favorable to the player, and those numbered 6 down to 2 as unfavorable. The middle cards—7, 8, 9—I determined as having a moderate effect on a player's chances, so I didn't worry about them. I concentrated on the two groups of high and low cards. In this way, all I had to do was keep track of the balance of favorable to unfavorable cards remaining in the deck. So as the cards went by, I didn't count them individually; I counted how many more were in one group than the other. With a little practice, I was able to do that very easily.

As the blackjack dealer went through one deck (or more—nowadays, some use up to eight decks), I would keep a mental tally of the cards in those two groups so I'd know, at any point, whether there were more high cards or more low cards left in the deck. That, of course, told me whether I was more likely to draw a high card that might push me over 21, or a low card that would not; and I was able to place my bets accordingly. This system gave me an advantage that produced some pleasant results when I played blackjack in Las Vegas.

I came up with this little mental exercise well before *Beat the Dealer* or any of the now-numerous self-help gambling books had been published. When I set myself the task of outwitting the casinos, they had no countermeasures against gamblers whose "luck" was too good to be good for business. Later, when they did begin to institute modern countermeasures, another of my innovations enabled me to take *counter*-countermeasures.

GETTING AHEAD BY USING YOUR HEAD

When I entered college, I was a good deal younger than most of the student body—which didn't do much for me with the coeds. But things started to turn around for me the day I outsmarted a machine.

At one of the college social clubs was a combination game and slot machine whose theme was horse racing. There were seven racehorses, and the idea was to pick the one that would come in first. Players bet on the horse of their choice by putting a nickel in its slot. Only after a bet was placed would the machine display the odds for that race. Typically, the odds were 2:1 or 4:1 and the like. The jackpot paid off 50:1.

After observing other students' play for about 45 minutes, I realized that the same odds seemed to be coming up in the same sequence as during the first half hour. I watched for another half hour or so, to confirm that the odds were repeated every 30 minutes.

Once I knew which odds would come up, winning was reduced to simple arithmetic. Whenever the odds were 8:1 or better, I'd bet on *every* horse (to be sure to win). That cost me seven nickels (35 cents). Eight-to-one odds paid 40 cents, for a 5-cent profit. Obviously, with higher odds, I'd clear even more. By knowing when favorable odds would come up, and then betting on every horse, I couldn't lose.

I decided to capitalize on my investment, so to speak: When I knew the 50:1 jackpot odds were coming up, I approached an attractive coed and asked her to pull the handle on the machine to start the race—"for good luck." (An older coed would do this, even for a "youngster" like me.)

Needless to say, when the jackpot odds appeared on the machine, the girl was delighted. She made a basket out of her skirt and into it poured the 50 nickels. Being my "lucky charm" was fun for her, and she became my date for the evening.

(continued)

Using my head to outsmart an early computer game gave me a real advantage in meeting college coeds.

For instance, dealers began to reshuffle the cards whenever a player increased his or her bet. This was done to prevent players from increasing their bets when the cards became favorable. I took advantage of their ploy: whenever the cards became *unfavorable*, I increased my bet, whereupon they reshuffled the cards and gave me a new deal. I was able to use the casinos' own tactics against them, the way a wrestler can sometimes maneuver so that his opponent's own weight brings him down.

Years later, at the Hudson Institute, I demonstrated the mind-over-machine power—again using cards—on a RAND computer that had been programmed to figure out how to win at blackjack. Herman Kahn, the president and founder of the Hudson Institute, a prestigious think tank, informed me that the computations the computer performed were the equivalent of 35,000 years of human labor. Totally unimpressed, I retorted that I had done the same analysis in a few hours over three weeks' time, using only a pencil and paper.

Naturally, he wondered how. I told him I had simply done my calculations using a number system with a base of 13 (again, because there are 13 cards in a suit). Thus, the numbers used in my calculations were the numbers of 13s, and the result determined the chance of winning with each play without having to use any multiplication or division. (Base 13 is a counting system of 13s, instead of the system based on 10s that we commonly use. In base 13, the sequence of numbers is as follows: 1, 2, 3, 4, 5, 6, 7, 8, 9, a, b, c, 11, etc. I did all my calculations in base 13, and only at the end converted the answers back to base 10. This vastly simplified the calculations.)

"If we had used base 13," Kahn exclaimed, "we could have done the work in two minutes!"

Resorting to an alternative system of numbers is another example of finding another way to a solution. And not only can finding another way lead to new solutions, it can also simplify your life and enable you to solve problems a lot faster.

Staying in Charge

To this point in this chapter, I've used some fun examples to make my point. Let me share with you here a more serious story that demonstrates how far superior the human brain is to any computer.

The supersonic Concorde has an exemplary safety record. (The recent accident on takeoff turns out to have been caused by debris left on the runway by another airplane, not any defect in the Concorde itself.) I was once invited to join the crew in the cockpit of one of these flying behemoths during its landing. Let me preface this story by telling you that the craft had automatic throttles for power controls. It did not have the clutch arrangement I had invented, which allows the pilot to override the automatic system. I was told that once the automatic system had been engaged, the pilot was "out of the picture" as far as the flying of the plane was concerned. The machine was, at that point, completely in charge.

I wasn't entirely comfortable with that attitude, because, as I've made clear in previous chapters, I'm a firm believer in the superiority of human intellect over even the most powerful computers. It turned out that our landing was about to prove me right.

After the descent was given over to the automatic control, when we were barely 200 feet from the ground, the automatic computer control, for unknown reasons, turned off one of the engines! Though we landed safely with one dead engine, the captain decided it shouldn't happen again. He was convinced, as I had been all along, that the system needed a mechanism to allow the pilot to override the automatic controls.

Minding the Mental Store

No doubt about it, computers are remarkable in many ways. They can hold massive quantities of information in what we anthropomorphically call their "memories," and they can process certain things very rapidly.

But they lack one critical capacity: they cannot be inventive. An inventorship chip, an imagination algorithm, and an innovation protocol for computers do not exist. They never will. Anyone interested in living life imaginatively, in solving problems inventively, should keep that in mind, along with these other inventorship points from this chapter:

- Because we have imagination, we can figure machines out, but they cannot do the same to us.
- To develop the inventorship mind-set requires stepping back from the world of machines and into our own heads.
- With a machine, if you create identical circumstances, you get identical results. Using your brain, you won't— and sometimes you'll get better results.

7

Inventorship and the
Entrepreneurial Spirit

The by-product is sometimes more valuable than the product.
—HAVELOCK ELLIS (1859–1939)

In the recent past, the business headlines have heralded numerous so-called *mega-mergers* of companies in all areas of commerce, from automobile manufacturers to banks to telephone companies. The trend to *big* is in, and it seems the big keep getting bigger. But at the same time—fueled by the computer technological developments of the past decade, which enable entrepreneurs to compete in the "land of the giants"—a concomitant trend of small business launches has also captured media attention.

Computers, as I said in the previous chapter, are only tools. The significant factor for companies that succeed as "small fish" in the "big pond" is inventiveness. Big companies have no monopoly on inventorship. In fact, their size can cause them to become overly bureaucratic or complacent—or both—and hence less willing and able to be as inventive as many smaller firms. Large groups are unwieldy by the least common denominator—management by committee. Inventiveness, innovation, daydreaming—all these are more characteristic of a single mind.

If you dream of starting your own small business, your success will not depend on the size of your company or the size of your bankroll; it will depend on the size of your thinking, along with the effort you invest in it. Many, perhaps most, small businesses that not only survive but *thrive* among the giants are able to do so because they incorporate inventorship into their business practices and goals. My own company, Safe Flight Instrument Corporation, for example, has prospered for more than 50 years in competition with firms many times larger, simply because we keep finding another way.

More important, although our products are not always the lowest-priced or the most aggressively marketed, they are usually the best and often the most innovative. Sometimes, we are the first to offer a product that fits a special need. For example, one aircraft required installation of a control stick shaker device to give pilots a tactile warning of impending stall. However, existing models did not fit in the space available on the aircraft. At Safe Flight Instrument Corporation, we designed one that operated in a coaxial manner. It encircled the control stick instead of being mounted *alongside* it and taking up more space.

Discounting Money

As I just mentioned, the size of your bankroll is not the most important factor when you launch a small business; yet many people believe it is, and that belief prevents them from advancing their plan. I have given lectures on starting businesses to MBA students at Columbia University, to working students at Westchester Community College, and to undergraduate students at Manhattanville College, and I always ask the students how many of them think they have enough capital to start a business. Out of a typical class of about 85, maybe one or two will raise their hands. But when I'm done speaking, I ask the question again. Every hand goes up.

If you have a desire for your own business, but have only a little money to put into it, think of it this way: you're in good company. Henry Ford, Bill Gates, Charles Goodyear, and countless others started out the same way. As Philo T. Farnsworth, who invented a device you may be familiar with—it's called television—said: "Important inventions are made by individuals, and almost invariably by individuals with very limited means."

Following in the footsteps of my father, who started his rubber cement company in the hayloft of a barn, I began Safe Flight Instrument Corporation with an investment of $14,000. Our first factory was a carriage house rented for $100 a month. Lack of capital isn't an obstacle to business success, unless you believe it is and allow that belief to keep you from getting started.

In some cases, not having a lot of money at the beginning can even be an advantage. Beginners with too much capital often don't know how to use it properly. Proof of this can be found in the numerous "dot-com" launches of the recent past that were funded on huge amounts of venture capital and failed dramatically.

My advice to those launching a business venture is:

1. Start small.
2. Try different things—that is, be innovative.
3. Make, and learn from, all of the inevitable mistakes.

In a small-scale operation without a lot of capital, the mistakes you make will teach you invaluable lessons; they won't destroy you and/or your business. Rather, those mistakes will make successful growth possible later on, when you are better able to handle it. When you can't solve a problem just by writing a check, you will learn to come up with a more inventive way of solving it. And new ways of doing things are precisely what make new businesses successful.

"Innovations," as management consultant and writer Peter F. Drucker says in his book, *Innovation and Entrepreneurship* (New York: HarperBusiness, 1993), "had better be capable of being started small, requiring at first little money, few people, and only a small and limited market." That, he explains, is because "innovations almost always need modification before they become successful," and "the necessary changes can be made only if the scale is small and the requirements for people and money fairly modest." In short, money isn't the cause of great businesses; it is the result.

Making Something Out of Nothing

There may be no better illustration of that truth than the following story of two men who bought a national business for *nothing.*

If you've had a Coke lately, it most likely came from a bottle or a can. But originally you could only get Coca-Cola from a soda fountain, where it was "made" on the spot from a condensed cola-flavored syrup mixed with soda water. Asa G. Candler was a Georgia druggist when he bought Coca-Cola for just $2,300. He then built it into a nationally distributed drink sold from the nation's drug store soda fountains. That's the way he thought it would always be served to customers. But as Harvard's Richard S. Tedlow tells it, in his book *New and Improved: The Story of Mass Marketing* (New York: Basic Books, 1990), a couple of Chattanooga lawyers had other ideas.

One of those lawyers, Benjamin Franklin Thomas, had become intrigued with the idea of bottling Coke. (While in Cuba during the Spanish-American War, he had seen a carbonated beverage being bottled.) Fellow lawyer Joseph Brown Whitehead was also a fan of the drink; but, more important to this story, he was also a passionate fan of baseball, and wished he could enjoy his favorite soft drink while watching his favorite sport. (See Figure 7.1.)

Figure 7.1 The desire to have a Coke at a ballgame was the motivation for one of the most innovative business ideas in the history of American business.

The two lawyers approached Candler with the idea of bottling Coke so that customers could drink it at home, at the ballpark, or anywhere else. The druggist predicted that such a business would fail, saying, "I have very little confidence in the bottling business." Nevertheless, in 1899, he agreed to sell to Thomas and Whitehead—for one dollar—a franchise to bottle and sell his drink throughout most of the country.

The new franchisees didn't have the funds needed to open a single bottling plant. But, to an inventive mind, such an obstacle is no problem. The attorneys simply subfranchised their rights to numerous regional bottlers. Within just 30 years, by 1928, bottled Coke was outselling the soda fountain version everywhere.

Keep in mind: No matter how entrenched it may appear to be, the status quo may eventually be replaced by another way. The inventor of that other way will be greatly rewarded.

Making Lemonade Out of Lemons

"When you are handed a lemon, make lemonade." In the realm of inventorship, this cliché has great meaning, as the following examples will illustrate.

Minding Waste

Across America, communities and individuals have become committed to the concept of recycling. No question about it, finding alternatives to throwing things out isn't good just for our environment; it can also be a very good way to practice inventorship.

Consider the detritus from making pipes for smoking. Did you know that the sweetness of pipe smoke has a great deal to do with the quality of the briar used to make the pipe? Well, it does: the densest, most expensive briar roots make the best pipes. And when a pipe is ground out from the briar, the by-product of the process is briar sawdust. To most people, it would be just dust under their feet. But when I saw it pile up at a pipe factory I visited once, I suspected it might have hidden value. So I gathered up some of the "waste" and mixed it with water-glass (sodium silicate, used commonly as a cement or adhesive), an inert substance with no flavor. I put the mixture into a mold and made some briar sawdust pipes. (See Figure 7.2.)

The result? A good smoking experience—not surprisingly, because I was using the same top-quality briar. The difference was that my pipes were made inexpensively from so-called

Figure 7.2 One product's waste may be the basis for a new product's development.

waste by-product. (I only made seven of these pipes—one for each day of the week!—and this product was not sold commercially.)

The real value in many products available today is not, after all, in the raw materials from which they're made, but in

---

FLOATING YOUR IDEAS

Don't be afraid to let your mind wander when it comes to ideas for products or businesses. You never know which one is going to "float"—literally.

I came up with an idea for a soap that floats in another way. We're all familiar with the bars of Ivory soap that float. Publicists made the Ivory name a household word many years ago by implying that it could do this because it's "99 and $^{44}/_{100}$ths percent pure." In fact, its "purity" has nothing to do with its floating capability, which is the result of how it's manufactured. It isn't milled and molded as most soaps are; the formula is poured into forms without removing the tiny air bubbles that occur during processing. Therefore, Ivory bars are porous, enabling them to float. True, that's a nice benefit in the tub; it saves a lot of groping around when you're trying to locate your soap under the water. On the other hand, once you've got it, you find that its porosity also makes it slippery and quick to dissolve.

I reasoned that, instead of those thousands of tiny air bubbles, why not insert a flotation device into a high-quality, hard-milled, long-lasting bar? I imagined a hollow, plastic, egglike capsule—perhaps even containing a toy prize like the ones in Cracker Jack. This idea, which, by the way, I patented, also has the advantage of eliminating the dreaded soap slivers: the center of the soap is not soap, so you use the entire bar.

the thinking that amplified the value of those materials in another way.

MINDING ANNOYANCE

Instead of minding annoyances—letting them "get your goat"—I submit that minding annoyances productively can lead to great ideas. In fact, practicing this form of inventorship has made some people extremely rich. Inventors, or those with the inventorship mind-set, have to learn to see the opportunity those petty nuisances may present, and then to act on what they see.

I recall reading somewhere about a gentleman who was living in London around the turn of the nineteenth century and who was particularly observant and thoughtful. Walking down the street one day, he found himself in the mood for a cigarette, but he had none with him. He had to walk quite a distance before he found a place that sold tobacco products. Most people would have forgotten the inconvenience once they finally lit up, but not this man. Disturbed that he had to go so far for an everyday item, he considered that other smokers in that neighborhood must have the same problem. In his day (before smoking was known to be a health hazard), there were many more smokers, and he speculated that if he opened a smoke shop in the vicinity he might do some pretty good business.

He opened his first shop, and after it began to do well, he wondered whether smokers in general would appreciate easier access to their smoking supplies.

He surveyed the city, found a number of prime locations (that is, without smoke shops), and opened a chain of tobacconist shops. Needless to say, he became a wealthy man, all because he turned enduring a nuisance into seeing a need, which

Figure 7.3 A lucrative business emerged from reacting to an inconvenience.

he subsequently found a way to meet, for himself and many others. (See Figure 7.3.)

STICKING TO IT

Perhaps the most well-known contemporary example of a great invention born of frustration is that of Art Fry, part-time choir director and full-time inventor. According to author Michael Gershman, in his book *Getting It Right the Second Time*

(Reading, MA: Addison-Wesley, 1990), Art Fry's greatest success came as a result of tribulations with his hymnal.

Fry liked to mark the hymns for the day by putting scraps of paper in the book, but any movement on his part, such as signaling the choir to stand, caused his bookmarks to fall out. What he needed, he realized, was something that would cling to the paper but could easily be removed without damage to the book.

Then an idea came to him. (I leave it to you to decide whether it was the product of an inventive mind or of divine

Figure 7.4 Initially regarded as a failure, the adhesive on Post-it Notes stuck around to become a big-time success.

intervention.) Fry, who worked for 3M Corporation, had recently reviewed a product "failure," a new adhesive. The product failed because when put to the test, it wasn't sticky. Perhaps, Fry thought, it might work in another way. Back at his office, Fry smeared some of the adhesive on a piece of paper and found that it would stick to another piece of paper, yet could easily be removed. It wasn't a permanent fixative, as most adhesives are expected to be. (See Figure 7.4.)

Well, no doubt you know how this story ends: Eventually, based on Fry's findings, 3M produced tiny pads of paper, coated on the back, along the top half-inch or so, with the adhesive. Post-it Notes hit the market in 1980 and, within four years, had become one of the most successful new product launches in the history of 3M—indeed, for any company.

Thinking Small to Succeed Big

When you're thinking about a new business, don't dismiss the little ideas or what others may regard as low-value products. A lot of big businesses have been built on small, inexpensive products that sell in huge quantities. Here are some tales of the big wisdom in thinking small:

- Probably the biggest small product to take the world by storm was even promoted with the words: "Think Small." Yes, I'm talking about the Volkswagen Beetle. Its petite size did not prevent it from surpassing Henry Ford's Model T and becoming the best-selling car of all time. And one more thing: The original VW Beetle sold for about a third less than most popular cars of its day. Recently, the Volkswagen company revived the beloved

KNOWING WHAT'S IN A NAME

Most successful start-ups are based on a great idea and are named later. That said, if you have a winning idea for a small business, but experience inventor's block when it comes to finding a focus point for it—the "single selling proposition" as the marketing people call it—there's a simple solution that has worked for me: Pick the name first.

When I address college students interested in starting their own businesses, I divide the class into groups to simulate companies. The first thing I ask these groups to do is to come up with a name for their business. I've found that focusing students on the name immediately after they have the idea behind the company is beneficial; it starts them thinking, not just about the name, but about the nature of what they are doing.

A name helps define a business, and what starts as a naming exercise rarely ends there. Given the naming assignment, my students get excited and come up with concrete plans that include some very good, and focused, ideas. A name makes something real.

Try this exercise yourself when you are stuck. I think you'll find that putting a name on your project or your business can inspire and energize you. This book, for example, started with the idea of the mind-set behind inventions and innovations; then came the name: "Inventorship." Once I had a concrete name for my idea, I found it much easier to actually begin writing the book.

"Bug" as the New Beetle, and the company can't make them fast enough to meet the demand.

- Some 60 years ago, inventor Gene Kulka had a couple of good ideas. The first was that it might be possible to build a radio that wouldn't need an antenna or a ground for reception. That idea grew up to be the portable transistor radio, ancestor of all the handheld electronic devices that are so popular today. But what made Kulka's fortune was his second idea: the plug embedded in fluorescent light tubes, which makes them so easy to install. This product, which few people have ever even thought about, and which originally sold for 16 cents apiece, made Kulka rich. (Kulka went into production with the inexpensive little endplugs simply as a way to amass enough capital to develop his radio idea, but the plugs were such a runaway success that he didn't pursue the radio idea. Fortunately, others did. I had a similar experience when I invented a new type of engine or "prime mover" that consisted of a centrifugal turbine. To get money to promote it, I went into the Stall Warning business, which became so successful that I never pursued the new engine type. Sometimes, with inventorship, the means to an end becomes an end in itself.)

- Perhaps you've heard of a gentleman named William Wrigley, who did pretty well selling packages of chewing gum for a penny.
- I know you've heard of an outfit called McDonald's that didn't do badly selling hamburgers for 15 cents.

The value of a product is in what it does, not how much you can charge for it or its component parts. (See Figure 7.5.) In my own case (specifically, my first invention, the Stall Warning device for airplanes), my first prototype was hardly impressive to look at. It consisted of a pair of electrical contacts on a hinged "tongue," or tab, of metal about the size of my finger. It was made using a pair of metal snips. To this tab

Figure 7.5 Many of the most successful consumer products are simple, small, or even invisible to users.

was added a bicycle buzzer. The entire mechanism was powered with flashlight batteries. When I tested it on the front of an airplane's wing, I knew that when the airflow over the wing reached a dangerous angle, the tab would move, hitting the contacts, setting off the buzzer, and alerting the pilot that his plane was approaching the dangerous condition of aerodynamic stall.

It wasn't a big device, or even a sophisticated design, but the idea behind it—advance warning of a dangerous condition—*was* a big one, and that's what counted. And although the early models of the Stall Warning Indicator were priced in 1946 at $17, Safe Flight Instrument Corporation has sold its descendants for as much as $35,000 each.

Keeping an Eye on the Prize

The lesson of this chapter is that, when it comes to inventorship, the issue is not glamour, or sophistication, or price; the issue is *true* value—the value of the idea to solve a problem, meet a demand, or improve the world is some small way. Never be afraid to think small—just be sure you think small in another way. Remember:

- Inventorship success doesn't depend on the size of your company or the heft of your bankroll. It depends on the scope of your thinking and the intensity of your effort.

- The inventive mind can make something valuable out of what may be seen as worthless by others. Learn to view both waste and nuisance as opportunities to practice inventorship.

- To realize an inventive dream, you don't always have to have a lot of money. When you can't solve a problem just by writing a check, you will learn to come up with a more inventive way to do so.

8

Innovation in Public Policy

Every great advance . . . has issued from a new audacity of imagination.

—JOHN DEWEY (1859–1952)

As I've emphasized throughout the book, inventorship as a field of study and as a life discipline encompasses much more than machines and technology. It can be applied to many areas of life. And perhaps no other area of life is more in need of inventive thinking than that of public policy. Policymakers will always be hampered by political considerations, but it seems to me that another problem is a serious lack of new, inventive, workable ideas.

Whether you're conservative, liberal, or straddling the fence, when it comes to applying inventorship to public issues, the watchword should be: "Whatever works." Inventors must never fall prey to what some have called the NIH—Not Invented Here—syndrome. This syndrome is marked by a premature dismissal of ideas solely because they did not originate from our own company, our own part of the country or world, our own business or profession, or our own political party—or even our own heads! True inventors recognize that gold is where they find it; they are open to good ideas from *any* source. The trick is to find a common denominator.

Finding a Common Denominator

There are those who believe that everything in our world is ultimately related to everything else. That may or may not be true, but what cannot be disputed is that it is almost always helpful to see connections between things, whether they are people, events, mechanisms, or phenomena—especially those that appear to have nothing in common with one another. This is another key to inventorship.

A good friend once asked me to give an hour-long speech at an assembly of 2,000 students, at his son's high school. The subject was welfare reform, a topic of great interest to me, but unfortunately not (as I learned) to them.

The school was in an affluent suburb where, to residents, welfare was little more than a political issue. More important, at the time, the Vietnam War was raging, and the draft number lottery had recently been announced. Understandably, these young people were more interested in the war, which threatened their safe and stable world, than in welfare, which did not.

Needless to say, I was facing an unresponsive audience. My first effort to engage these students was to announce that I would speak for just 20 minutes and then take questions for the rest of the hour. The first question I was asked laid down the gauntlet. What, the student wanted to know, did this talk on welfare "have to do with the war?"

I knew I had to find some way to make what was happening thousands of miles away relevant to what was happening much closer to home—even though they didn't see it as closer to *their* homes. I told the students that we were fighting in South Vietnam because we wanted our system—democracy and capitalism—to prevail in that Southeast Asian nation. The North Vietnamese were fighting to have their system—autocratic Communism—prevail. I followed this by saying we were missing the point by fighting a war for this purpose. At that, 2,000 pairs of ears started to perk up. In faulting our system, I noted, the Communists have always pointed to our poor. Suppose, I ventured, we used the right kind of welfare reform to make our poor better off than anywhere else in the world. If we could reform our system's treatment of the poor, wouldn't the world take notice; and wouldn't the nations of Southeast Asia—and everywhere else, for that matter—adopt our system gladly? Isn't an effective political and economic system a more powerful way than force of arms to convince the world that democracy is better than communism?

With those comments, I had succeeded in changing the attitude in the auditorium. The second question, instead of attempting to put me on the defensive, was proactive: "How can we make that happen?" I began to explain the answer to that in some detail, and before I knew it, the bell rang to signal the end of the assembly. But no one moved! We all stayed for a second hour, discussing ways to accomplish our nation's objectives. (See Figure 8.1.) Throughout history, war and poverty have often been negatively linked, but, in this instance, the connection was positive. It provided a common denominator that enabled some very bright, but somewhat insular kids to take a broader view and gain new insights.

Figure 8.1   Most chasms can be bridged by finding the common denominator.

Giving Credit Where Credit Is Due

Based on my philosophy of doing whatever works and finding a common denominator, I also proposed what I called the National Tax Rebate. The details of that proposal are beyond the scope of this book, but I'd like to tell you just enough here to make some points about inventorship. [For those who are interested, my book *The National Tax Rebate: A New America with Less Government* (Washington, DC: Regnery Publishing, Inc., 1998) explains this plan in full.]

For years, liberals have warned us about the growing problem of poverty in our land. From conservatives, we've more often heard that our taxes are too high. In my opinion, both sides are right. Put another way, when I began to evaluate the problems of poverty and high taxes, I found them to be two sides of the same coin. I reasoned, therefore, that they must also have a common solution. I called my solution the National Tax Rebate. In a nutshell, the plan involves cashing out hundreds of inefficient and mistargeted government programs designed to help the poor or to offer financial relief to different groups of taxpayers *indirectly*, and redistributing the funds realized in this way by means of a *single* program that would deliver the money *directly* by both giving the poor a helping hand instead of a handout, and providing tax relief to the overburdened middle class. (See Figure 8.2.)

This direct approach eliminates the rules, the paperwork, and the bureaucracies, to give financial relief to the people who need it, whether they're poor people needing sustenance or middle-class people needing tax relief. The National Tax Rebate would be paid for with money we're already spending on all the "band-aid" solutions we've designed over the years, many of which have caused more problems than they've solved.

Figure 8.2   The National Tax Rebate would go to all Americans and would give the average family of four a total of $12,000 additional income each year.

Computer simulations done by the Institute for Socio-Economic Studies (the think tank I founded and head) with the Columbia University School of Social Work demonstrate that even if we eliminated all our present social and corporate welfare programs to fund such a National Tax Rebate, we'd have much less poverty than we do today. And if, year after year, every family of four received an annual National Tax Rebate check amounting to $12,000, we'd also have much happier middle-class taxpayers.

The Institute is actively promoting the idea of a National Tax Rebate, and the concept has received significant media attention, although, to date, it has not gotten much play on Capitol Hill. Like all good ideas, though, its time, I'm sure, will come. The plan is simple, but it would work. And, more important to the objective of this book, the plan shows that inventorship can be applied to any sort of situation.

Caring for All

Might inventorship also hold the key to solving one of the most critical and most hotly debated social issues of the past decade: affordable health care? I believe that it does. I have proposed a solution called Patient-Managed Care. As with my National Tax Rebate plan, the specific details of this solution are not important for the purposes of this book. What is important is how I arrived at the basic concept, for that process illustrates how inventorship can be applied to a seemingly intractable problem. It's also a pertinent example of how a solution to a problem can come from someone not an "expert" in the field.

It hardly needs to be said that managed care is not working the way it was anticipated and hoped. Almost no one, whether as a patient or a health care provider, who has had to negotiate and navigate these so-called health maintenance plans has come away unscathed—at the very least, the results are anger and frustration.

In brief, as we all know, managed care simply means that someone is designated to manage or make decisions as to which health care services a patient receives. The goal is to treat his or her current ailment while eliminating waste and "overtreatment" and preventing more serious problems down the road that will be even more difficult and more costly to handle. And, in principle, who can argue with those goals? The critical question is: Who does the paying and, therefore, the managing? Under these plans, it is usually an insurance company. But we all know the problems—in some cases, life-threatening—this has caused. In earlier managed care plans—specifically, Medicare and Medicaid—the government has done the managing. But that has led to out-of-control spending and arbitrary rules indicating the kinds of care that would be paid for.

In contrast, when most of us need food, housing, clothing, or any of the other necessities of life, we don't go to an insurance company or to the government. We pay our own money and make our own decisions. We're the managers, in other words, and, except for the poor, who are always a special case, we somehow get the necessities we need at a price we can pay. Why, I thought, can't that also work in health care? Why not maintain the very sensible idea of eliminating wasteful spending and outside care management? In short, why not just change the managers? Instead of insurance companies or the government, let's make individuals or families the managers, then give them the incentive to manage well. Who knows more about what they need than the people themselves? If people are qualified to manage their own housing, sustenance, child care, and other vital aspects of their lives, why are they regarded as incompetent to make decisions on their health care?

Under my Patient-Managed Care plan, the patients would become the managers, as they were before both managed care and government health care were implemented. (See Figure 8.3.) The plan acknowledges the high cost of medical care by offering subsidies to help people who are unable to work while they are sick, but it also encourages them not to be wasteful in their health spending. It allows people to get whatever kind of care they need from whatever provider they choose, whatever it costs. But it also provides reasonable incentives both to patients and to physicians and hospitals to control expenses.

The Patient-Managed Care plan would provide universal catastrophic medical protection. Out-of-pocket medical costs

Figure 8.3 Under Patient-Managed Care individuals and families would pay their own medical bills, up to a certain percentage of their income; then insurance, with a small copayment, would take over.

would be limited to a percentage of each person's income, and people would be responsible for all costs up to their personal limit. This plan would require NO new expenditures, as funding would be rolled over from existing health care spending sources.

As with the National Tax Rebate concept, my Patient-Managed Care plan is being researched, developed, and further refined at the Institute for SocioEconomic Studies, which cosponsors conferences on health care with other institutions as part of its initiative to promote access to quality health care for all Americans.

I think my Patient-Managed Care plan has the potential to straighten out an overly complicated industry by using a very simple idea that, essentially, adapts a system that works in other aspects of our lives. It takes a problem and makes it the solution. It is inventorship at work

Making New Combinations

Inventorship applied to public policy sometimes consists of combining familiar ideas or elements to generate a new idea. In the days before the advent of federally funded student loans, I came up with another way of managing student scholarships. Traditionally, a student who is awarded a scholarship or fellowship based on merit is given money for his or her education, and that money does not have to be repaid. In contrast, a student loan has nothing to do with merit, and the money does have to be paid back.

My idea was a hybrid I called Chain Scholarships, named after the chain letter idea. Like conventional scholarships, they were awarded based on merit to students of demonstrated character; like a loan, the money was to be repaid after the student graduated and got a job. But unlike the usual loan, repayment was on the honor system; that is, there was only a moral—not any legal—obligation to repay. (As such, the repayments were charitable contributions.)

The Chain Scholarship plan was an example of how two procedures can be united to form an innovative third, and it worked exceptionally well. Though no interest was charged, many students paid back more than the amount of their grant, and the repayments went to fund grants for other students.

GIVING CANCER PATIENTS A LIFT

After the Chain Scholarships had been replaced by the federal student loan program, in a way, Chain itself became part of a chain—enabling the start-up of another nonprofit organization, which benefits many people today, and is helping more every year.

My friends Jay Weinberg and Priscilla ("Pat") Blum and I had all had some personal experience with cancer and knew that patients often have difficulty, whether financial or physical, in traveling the great distances required, in many cases, for optimum treatment. We also knew that many corporate aircraft fly on business trips every day with empty seats. Pat asked us if we could develop the idea of asking corporations to accept such patients as guest passengers, and Jay and I got to work.

To launch our new endeavor, I provided the Chain Scholarship's corporate foundation, along with its assets. That became the Corporate Angel Network, Inc. (CAN). Its mission is to arrange air transportation free of charge for cancer patients traveling to and from recognized treatment centers, by making available to them empty seats on corporate aircraft operating on business flights.

On December 22, 1981, I flew CAN's inaugural flight, between White Plains, New York, and Detroit, Michigan, transporting the first patient, an adolescent boy, home for Christmas after treatment. Since then, CAN has arranged more than 13,000 flights and is still growing.

CAN has received many awards in recognition of its service to cancer patients, including the *Volunteer Action Award,* the highest volunteer award from the President of the United States. Currently, 60 volunteers and a small paid staff coordinate the flights provided through the cooperation of over 550 major corporations.

CAN's value to cancer patients is incalculable. They save money by avoiding commercial travel, and in the quiet environment of a

(continued)

private plane, they travel undisturbed by the crowds and congestion associated with high-density airports. Patients fly in comfort and dignity, cheered by knowing that somebody cares enough to help them during a stressful time in their lives.

And the CAN experience is not a one-way street. Patients appreciate the help, of course, but corporate executives thank CAN, too, for creating an opportunity for them to show their public interest. Volunteers at the CAN offices (many of whom are retired executives themselves) feel a serious sense of dedication and accomplishment in their work, as do the corporate schedulers they work with to arrange the flights. CAN is truly a win-win situation for everyone involved. It is a wonderful example of what an innovative program, with cooperation and community involvement, can do.

All it takes is good ideas, good luck, and good will! If you look around you with an inventorship mind-set, you may find that there's something close to home that you can do to make life better for others—and more rewarding and interesting for yourself.

Before being phased out when the government launched its own program, this inventive combination of two well-proven ideas constituted a successful new approach to financing higher education. Over 500 students in the Chain program received college degrees.

Taking an Interest

It's easy to complain about how badly we think our policy-makers are addressing the issues of greatest import to us—

taxation, health care, education. It's more difficult, but also more rewarding, to consider how we all might be able to do something much more proactive than just voting for this person or that. No, we don't all have access to the top leaders of government, but, as I've said throughout the book, some of the best ideas start small and simply. Grass roots policymaking can eventually be adopted on a larger scale. Why not, I say, apply inventorship at the neighborhood or community level? You never know where it might lead.

- Inventors must avoid falling prey to the Not Invented Here syndrome.
- Look for connections between people, events, mechanisms, or phenomena, for therein often lies the solution.

9

Commitment: The Key to Successful Inventorship

Genius is one percent inspiration and ninety-nine percent perspiration.

—THOMAS EDISON (1847–1922)

We all know the cliché "Use it or lose it." It's a colloquial way of saying that human capabilities are strengthened only when we exercise them. Inventorship is no different. The sooner you begin to approach life from the inventive point of view, the greater, more frequent, and long-lasting the rewards will be. I know: I am more than 80 years old and still inventing—if anything, at a somewhat faster pace than before.

My latest design concept is the world's largest helicopter. With 200,000 pounds of lift—double the power of the previous largest aircraft of this type—it will be able to do things never before possible, such as hold a gap-closing girder in place while it is riveted, or put a derrick on an ocean platform.

I am also working on:

- A new zero-visibility aircraft-landing device.

- An aircraft "in-the-air" separation system.

- A system that will allow aircraft to reach a desired altitude faster, farther down course, and to consume less fuel.

I'm not the only one to enjoy the lifelong benefits of the inventorship way of life. I recently received a letter from a fellow inventor who is also in his 80s. He shared with me his idea for a tornado warning system he has patented, and it sounds promising.

According to the National Weather Service, people have only 10 seconds to take cover before a tornado strikes. That's not a lot of time. But this inventor has proposed a device, similar to a smoke alarm, that can be installed in homes. Here's how it would work. Low barometric pressure is the traditional warning sign of an impending tornado. His invention would alert people to a sudden drop in atmospheric pressure, which is what occurs *immediately before* the drop in barometric pressure. The alarm, in other words, would be triggered not by low pressure, but by detecting an increase in the rate of the lowering of the pressure. This could provide a full 60 seconds of warning time, instead of 10 seconds. That extra 50 seconds will save lives.

We hear a lot of talk about how many of the elderly suffer memory loss; and of course that's true: many do. But inventorship does not involve memory; it's not about what was, it's about what has never been.

I recommend that you develop your inventorship capability now. If you do, I can assure you a very exciting old age. You might not be able to get a job as a sky diving instructor, or train for the Olympics, but you might be able to get a patent (the U.S. Patent Office doesn't ask your age), and maybe even save a few lives. That sounds at least as satisfying as anything you might accomplish on the golf course.

As I said in Chapter 2, "The Age of Innovation," inventorship comes naturally to us when we are children. If we don't suppress it, it can last as long as we do. The objective of this chapter is to give you some guidelines for making that happen.

Practice Makes Inventorship

Let me quote another oft-repeated cliché: "How do you get to Carnegie Hall?" "Practice, practice, practice." Well, I'm here to tell you the road's the same to the Inventors Hall of Fame. But the point I want to emphasize in this section is the message behind the message. An inevitable by-product of practicing inventorship, of exercising the inventive mind, is failure. Too often, that truth keeps people from trying things another way or taking the road less traveled. Not all new ideas are good ones, even those born from the most brilliant minds. More important is to recognize that worthwhile innovations often emerge from ideas that, in their original form, didn't quite cut the mustard.

Author Michael Michalko reported ("Thinking Like a Genius: Eight Strategies Used by the Supercreative, from Aristotle and Leonardo to Einstein and Edison," *The Futurist*, May 1998) on a study of more than 2,000 scientists conducted by Dean Keith Simonton at the University of California at Davis. Simonton found that even the most capable scientists produced

RISING ABOVE

Here's a good example of not being afraid to fail. At Grumman Aviation in 1941, a colleague told me of his idea for installing a propeller on the roof of a car, the advantage—as he imagined it—being that when caught in traffic, drivers could just turn on the propeller, and, in effect, "rise above" the bottleneck. He was

Behind every successful innovative idea are a great many more unsuccessful innovative ideas.

convinced that Grumman should make his so-called propeller car. What he failed to address was the resulting air-traffic jam if everyone else with propeller cars took off at the same time. This was not a good idea—but I wonder whether he kept trying and eventually succeeded in inventing something else.

more than their share of duds in the innovation department. Eventually, however, these scientists came up with "the goods," in part as a result of first learning what didn't or couldn't work.

Let me paraphrase a third cliché: If it's worth doing, it's worth doing poorly—and continuing to try until you get it right! Remember my telling you about the time I lectured to 200 scientists at the Goodyear Tire and Rubber Corporation in Akron, Ohio? In my speech, entitled "How to Be Innovative," I told them that was my one general rule for stimulating innovation in any field.

Thinking Beyond a Solution

Have you ever asked yourself, in frustration, "Why is it that I always find what I'm looking for in the last place I look?" The answer, of course, is that once you've found it, you stop looking, so of course it's in the last place you look. My intent is not to be glib; there's a pretty good lesson here. It makes sense to stop looking when you've found something such as your car keys or your sunglasses. But when you're looking for an innovative solution to a problem, you should never stop looking.

When you've found your missing keys, there is no doubt you've found exactly what you need. But when you find a solution to a problem, there are no such guarantees. All you know is that you've found *a* solution. You don't know that you've found the *only* or the *best* solution. When we find something that works, we tend to assume it's the only thing that will work. We stop looking for alternative solutions, and that's contrary to the inventorship way. There is usually not just one way to solve a problem or even one way to look at a problem. And the first solution we find may not be the best.

Most of us don't marry the first person we go out with or buy the first car we test-drive. Why should we be content with the first solution we arrive at to take care of our problems?

Finding a solution—especially to a daunting problem—can be exciting, rewarding, even a relief. But I'm suggesting that instead of saying, essentially, "Phew! That's the answer! I'm done here," a better response is, "Great—that's one way to solve this." Then, go back to square one, and think about the problem in a different way. When you do this as a matter of

Figure 9.1   After finding a needle in a haystack, keep looking for the rest of them.

practice, you may come away with a second—or third or fourth—option that has a good chance of being better than the first. Einstein once said that whereas most people were happy to find the one proverbial needle in a haystack, he preferred to keep searching until he had found all the needles. (See Figure 9.1.) Spoken like a true inventor.

Confronting Issues Head-On

Making a commitment to inventorship also enables you to face challenges more confidently. This has two immediately apparent advantages: (1) you learn that most challenges look a lot tougher to deal with than they actually are; and (2) because many others back away from challenges, you will often have very little competition, and for that reason alone will have a much better chance of succeeding. A can-do attitude will stand you in good stead whether you are trying to solve a mechanical problem or just confronting a wrongdoing. Let me give you an example.

Many years ago, my wife and I were enjoying some downtime at the Hotel Greenbriar in West Virginia. One evening, we decided to go out to sample the area's entertainment. But we were advised not to frequent a particular local club, because the gaming was reportedly crooked. That sounded like an interesting challenge, so that's where we decided to go. I wanted to see if I could determine how the club was cheating its clientele. I decided to make it easy for them to cheat me, so that they could make it easy for me to see how they operated. To that end, I asked my wife to don her fur jacket and emerald brooch, and I ordered a limousine to take us there. When our driver opened the car door in front of the club, we stepped out in style, and, as the saying goes, they saw us coming.

I went to the cashier's window and purchased a stack of $5 chips for my wife and a stack of $25 chips for myself. It didn't take long to confirm that the rumors about the place were true. At the first blackjack table we went to, I could tell that the dealer's eyeshade was polarized, which enabled him to detect marked cards from their backsides. Another giveaway was his "mechanic's grip." (See Figure 9.2.) This special way of holding a deck allows the dealer to deal the second, rather than the top card, if, thanks to the polarized eyeshade, he can see, for example, that the top card is an ace. [For more details on this topic, see *Danger in the Cards* by Michael MacDougall (Chicago: Ziff-Davis Publishing Co., 1943).]

My wife began playing with her $5 chips, losing every hand until half her money was gone, at which point she urged me to

Figure 9.2   Confronting challenges head-on can help you "see" your way to a solution.

get into the game. I declined; at the same time, I conspicuously fingered my $25 chips and said, in a stage whisper, "The deck looks cold." That caught the dealer's attention. Shortly after my comment, my wife began to win as consistently as she had been losing earlier. The dealer's intent was to lure me, with my $25 chips, into the game. I continued to resist, even as my wife kept winning. When she had doubled her money, she stopped playing. I cashed in her chips, as well as all my unplayed chips, and we went back to our hotel. Though the game was rigged against us, I had fun finding another way to win.

I'm not advocating that you patronize disreputable establishments in order to prove you're more inventive than they are. What I am suggesting is that you not back down from daunting challenges, because, by evoking inventorship, they can often be the most rewarding.

Making Fantasy Reality

No matter how technologically advanced we become, the future remains humankind's greatest mystery. Yet visionaries of every era attempt to predict it. The result is that, sometimes, they seem to lead the way there.

One visionary I greatly admire is Hugo Gernsback. When he was president of the Electrical Society of the United States, some 75 years ago, he wrote a science fiction book titled *Ralph 124C 41+* (Boston, MA: Stratford, 1925). In this fantasy, he described inventions he foresaw for the year 2660. Gernsback's imaginings turned out to be anything but fantasy. As far-fetched as his turn-of-the-last-century dreams seemed to be at the time he wrote the book, humans have already gone beyond most of what he imagined.

In *Ralph 124C 41+*, Gernsback described radar, microwave technology, even space travel. He actually named "television," and described it as being carried into our homes via cables; he also envisioned people carrying embossed credit cards in their wallets. Even his title was on target, for he prophesied that, in the future, people would be identified by numbers. (I think we all know how close we are to making that prediction come true.)

(Note: As a tribute to Gernsback, a pioneer of the genre, the premier award for science fiction literature is called "the Hugo.")

How could Gernsback—and others like him—see the future so clearly? That's the gift of inventorship. It lets us know, or at least imagine, the unknowable. Inventorship opens the door and shows us the way—another way.

As we head into the final chapter of this book, which is all about living the inventive life, I remind you of the following inventorship points from this chapter:

- Worthwhile innovations often emerge from ideas that, in their original form, were failures.

- When you find a solution to a problem, don't stop looking; often, the most innovative and best solution will be your second, or third, or fifth!

- Making a commitment to inventorship enables you to face challenges of all sorts with greater confidence.

- Inventorship, once developed, is a "friend" for life.

- Inventorship is age-proof. We may lose some of our physical capabilities as the years go by, but we don't lose our imagination. Even the oldest among us have dreams, and dreams are the raw material of invention.

10

The Inventive Life

Imagination has brought mankind through the dark ages to its present state of civilization. Imagination led Columbus to discover America. Imagination led Franklin to discover electricity. Imagination has given us the steam engine, the telephone, and the automobile, for these things had to be dreamed of before they became realities . . . The imaginative child will become the imaginative man or woman most apt to invent, and therefore to foster civilization.

—L. FRANK BAUM (1856–1919)

As we have seen, inventorship is a powerful tool we can use to make our machines more capable, our work more productive, and our businesses more successful. In these and other less dramatic ways, inventorship transforms our lives. To conclude this book, I want to tell you how living the inventive life—as opposed to working as an inventor—has shaped my world and that of the people I share it with.

Throughout, I think I've made it clear that inventorship is not just something I employ during working hours. To me, it is a way of life. My family and I have lived and continue to live more adventurous, entertaining, and rewarding lives because of inventorship. Although I have received more than 100 patents, most of the ways I have applied inventorship throughout my life have led to innovations never filed with the U.S. Patent Office.

The most important lesson of this chapter—indeed, of the book—is that you can apply inventorship to your job, your relationships, your favorite sport, your social life, your hobbies. Once you have learned to practice inventorship, it's like riding a bicycle. You never forget how; in fact, it becomes second nature. Inventorship can be as effortless as daydreaming, but a lot more productive.

Inventing a Life

One of the best features of inventorship is that it's portable; you can take it anywhere. You can practice inventorship in a laboratory, workshop, or office, but you can also engage in it in the kitchen, on the golf course, in church, in the classroom, at a party—in short, anywhere, even while you're asleep, as I described in Chapter 3, "The Rule of Inventorship."

SHARING SIGHT

On July 16, 1969, even as the sun was still rising, thousands of people were already lining the highways and crowding the beaches around Cape Canaveral; at the same time, all across America, millions more were turning on their television sets, while their coffee was still brewing. For on that day, the United States and NASA were launching Apollo 11, aboard which, if all went well, were the first men who would walk on the moon.

As the viewing stands began to fill with government officials, heads of state, and other VIPs, one small group—two middle-aged men, one attractive woman, and a big German Shepherd—was making its way through the crowds. I was one of the men; the good-looking woman was my wife. The other man was my friend Brian Wallach, and the dog, Max, was his.

As we were walking, I found myself recalling the sequence of events that ended in our being among the privileged few allowed inside the Space Center on that historic day. It was no mystery why I was there. Thanks to my work in aerospace and aeronautics, I had long-standing links to NASA; in addition, I had worked with Neil Armstrong years earlier. Perhaps most pertinent, my old friend Jerome Lederer was then NASA's Chief of Air and Space Safety. I wasted no time in accepting the invitation to attend the launch.

If there was one cloud over the upcoming event, it was that, at the time, NASA was the focal point of a great deal of criticism regarding the amount of money being spent on the space program. Many Americans thought that the $8 billion it was costing to give Neil Armstrong and his colleagues a stroll on the moon could be better spent a little closer to home. As the United States accelerated its "race for space" with the then Soviet Union, there was a growing feeling that our nation was sacrificing its social conscience to technology.

It seemed to me that NASA was in need of something that would refocus public attention on the upcoming marvel, at least for the weeks preceding the event. In my experience, nothing captures public attention like a good human interest story, so I suggested to the NASA officials that they include, among the select few being invited to view the launch, someone who could not actually "see" the history-making liftoff of Apollo 11—that is, someone who was blind. It took them a month to agree. Then they said they didn't have anyone in mind for this honor, so I nominated Brian Wallach, a member of the Institute for Socio-Economic Studies, the think tank I founded in White Plains.

Though blind since childhood, Brian hadn't allowed his disability to limit him in any way. He was an insurance underwriter by profession, and was active in civic and municipal affairs. As president of the White Plains Beautification Society, he campaigned for a carillon in the downtown area, and went on trips to "see" what other beautification groups were accomplishing. He was willing to try just about anything, even skiing, which he enjoyed every winter, finding his way down the slopes safely by following the sound of a bell worn by the skier ahead of him.

He and I had many adventures together. In an empty parking lot, I taught him to drive. And he sailed with me frequently. One blistering hot day, in the middle of Long Island Sound, I said to him, "Want to take a swim?" "Sure," he answered, and without a moment's hesitation, jumped over the side. And there was Brian, swimming along by the boat, miles from shore. Most people, sighted or unsighted, myself included, probably would have been a lot more cautious.

I also gave Brian flying lessons; I had a twin Bonanza then, with dual controls. At the time, one of Brian's many extracurricular interests was radio. He was doing some announcing for WFAS, a local news/talk station, and he broadcast a kind of "You Are There" account of his first experience at the controls of an airplane, during which he described the wonderful sensation and the "gorgeous view of the Palisades." This was not to be Brian's last broadcast about his personal involvement in aeronautics and aerospace.

Naturally, Brian jumped at the chance to take his place at the Kennedy Space Center for the Apollo 11 launch. He reminded me that he'd have to take Max, his seeing-eye dog. I relayed this information to NASA, and told them Brian would also need a *human* companion. In this way, Max and my wife were cleared to attend.

On launch day, Brian decided that the stands, now rapidly filling with reporters and celebrities, would be too difficult for Max to navigate, so our little group found a spot on the ground in front of the blockhouse, a sort of bunker from which the flight would be controlled and monitored. One by one, three old friends spotted us at our unusual vantage point and came over to join our party: Tom Watson, President of IBM; Jerry Lederer, who had helped me persuade his colleagues at NASA to invite Brian; and U.S. Congressman Ogden Reid from New York State's Westchester County.

Together, we engaged in inventorship as we shared our sight with Brian. My wife described the prelaunch scene in vivid detail. It was a glorious day, brilliantly blue and clear, and already getting hot, as only Florida in July can. Setting the scene for Brian, she told him about the crowds, the launch pad, and the gantry crane and other structures supporting the spacecraft.

Our position in front of the blockhouse put us as close to the launch pad as we could possibly be without having to be inside a protective enclosure. At launch time, none of us needed our eyes to experience the enormity of the event. It was all about feeling: the pressure waves from the blastoff were overpowering as they vibrated and slammed against us. It felt as if the Earth itself was shaking. It was like being on a gigantic jackhammer; it shook our rib cages. Every person there was screaming maximum volume with the excitement and the thrill of it. Remarkably, Max sat quietly next to Brian and didn't budge; he didn't even seem nervous. (See Figure 10.1.)

Figure 10.1   Sharing with others is an effective and generous way to practice inventorship.

After Apollo 11 had lifted away from the gantry and moved off into the atmosphere, things quieted enough so that we could resume our "sight-sharing" with Brian. Tom drew Brian a word-picture of the ignition, the ascension into the gantry, and the liftoff. I described the stage separation, the flight Mach numbers, the Max Q spot (the point at which airspeed goes up and air density goes down), and the changeover from vertical climb to horizontal flight. As we spoke, within just minutes, Apollo 11 was little more than a speck, curving out of sight. Soon, all that remained was a wisp of condensation, like a puff of cigar smoke dissipating in the air.

Fortunately, everything had gone according to plan for NASA. And once reporters had all they needed for their headline story about the launch, a number began queuing up to talk

SEEING THE TANK HALF-FULL, NOT HALF-EMPTY

It didn't surprise me that an event as momentous as the Apollo 11 launch would get my inventorship juices flowing. One aspect of the takeoff process in particular stuck in my mind: Spacecraft rockets

Those living the inventive life pass up no opportunity to apply inventorship—even to a space launch.

(continued)

SEEING THE TANK HALF-FULL, NOT HALF-EMPTY (CONTINUED)

consume more fuel just getting off the launch pad than they take into space with them for the entire rest of the trip. This means that they begin their journey through space with a tank that's more than half empty. Imagine owning a car that burned more gas just getting out of the driveway than it did in the next thousand miles of driving. That pretty much describes the way spacecraft operate.

That seemed wasteful to me, to say the least. It made more sense for these spacecraft to start out with a full tank of gas, and thus be able to go further into space. My idea to make this possible was, essentially, to "fill 'er up" a second time at the top of the launch tower through an umbilical cord-like device.

Though NASA was interested, they never followed up on my idea—probably because pumping the fuel that way might lead to an explosion. Perhaps someday someone will find a way to safely accomplish the second fill-up—or, better yet, a way to get off the pad without burning so much fuel.

to us, intrigued that a blind person had been included among the launch viewers. To each one who asked "What's a blind person doing here?" I answered, "Blind people are people, too." Eventually, tired of the same question and my same answer, I took a more inventive approach to address their incomprehension. I asked one of the reporters, "Can you keep something confidential?" "Of course," he replied. "NASA is thinking ahead," I told him. "It has plans to go to other planets. The next one is Venus. Now, you know that Venus is covered with a dense fog. And the best person to handle these conditions is a blind person!" Within a few seconds, the embarrassed reporter realized I was pulling his leg and made his retreat. But who knows . . . .

Brian later taped his "observations" of the launch for a radio broadcast, and it became a very popular version of the event. As a blind person, he was expert at describing events and experiences to "blind" radio audiences—for, while listening to the radio, we are all without sight and must use our imagination to visualize a scene.

SHARING COLOR

While you don't have to be able to see to experience a space launch, you do need pretty good vision to become a pilot, and you must be able to distinguish color. At night, pilots have to be able to tell red from green because control towers use red and green lights as emergency signals after dark.

Inventorship enabled me to share sight—of color—with my wife. After she married me, my wife decided that she, too, wanted to learn to fly and, eventually, to become a licensed pilot. Initially, however, her pilot's license had to be restricted; she was barred from flying at night because she is color-blind. (In the test for color-blindness, the applicant looks at a graphic containing letters and numbers that can be discerned only by their contrasting colors. To a person who cannot tell red from green, the characters are invisible.)

The optical industry was not beating a path to my door in hopes of my discovering a way to correct color-blindness, but I had a more important customer. Basing my solution on the red and blue cardboard glasses popular in the 1950s for watching the 3-D movies or reading the 3-D comic books that were then all the rage, I constructed a pair of spectacles with lenses that were different colors. The right lens was tinted red, and the left was a neutral gray, which admitted about the same amount of light as the red lens.

When worn, the different-colored lenses caused each eye to have a different perception of colors—enough to make it possible to distinguish red from green. Alternately shutting each eye, my wife could now distinguish between the two colors. (See Figure 10.2.)

Wearing the glasses, she retook the test, passed, and was issued a license allowing night flying—of course, on the condition that she wear the "color glasses." The FAA did not care that my wife used another way to achieve her goal.

Figure 10.2 Using inventorship tactics, you can enhance other people's lives in meaningful ways. The red-gray glasses enabled my wife to pass her pilot's examination.

TAKING RISKS: PART OF INVENTORSHIP

No matter where your inventorship travels take you, if you're doing it right, you will have to take some risks. As an inventor who has spent his life designing and manufacturing flight safety instrumentation, I have more than once found myself in a risky situation.

One undertaking in particular, while flight-testing my company's Landing Speed Indicator aboard the U.S. Air Force's B-47 high-altitude jet bomber, turned out to be a lot more exciting than I expected. So much so that, as soon as it was over, I wrote a humorous account of the experience, to share with my wife. And now, some 40 years later, I'll share it here with you.

My tale begins at Wright-Patterson Air Force Base, "where our Uncle Samuel had just placed $15 million worth of airplanes at my disposal—three B-47 jet bombers. The purpose of this mission was for yours truly to go up in these monsters and see if his Landing Speed Indicator worked. If it did, we were going to be able to land these enormous machines at three times the speed of an express train on a 1½-mile runway. If not. . . . "

How do you fly a B-47? You don't. "She" flies you—if, when, and how she pleases. As I wrote, "The handwriting was on the wall, in the form of a mimeographed sheet issued by our same Uncle Samuel. This you must sign without reading it, or you will never go up in Miss B-47. [Once signed], it promises that your heirs will make no claims against Uncle if Miss B-47 should take you apart into several pieces."

My flying gear included not only a parachute, but also "a crash helmet, sewn-in oxygen tank, regulators, gauges, microphone, earphones, electric blanket, shoulder harness, crash belt, first aid kit, sunshade, and can opener. (I was suspicious of the can opener.)"

The preflight briefing consisted of little more than instructions on how to bail out of the airplane: "If the emergency siren wails three times, pull the emergency cabin depressurizer, pull the emergency air lock door, pull the emergency exit wind screen flap,

(continued)

fasten the emergency oxygen bottle, unfasten the emergency parachute release, and JUMP. Upon reading these helpful hints, the co-pilot muttered, 'Not me, brother. Every dummy [pilot] they've tried to parachute out of this plane so far has had its head sheared off when it was blown against the door. What I recommend is, crawl behind the seat and cover your face.' I guess this made sense. At least you wouldn't see what hit you."

Just after I was briefed and ready to go, fire engines came wailing up. "Apparently Miss B-47 resented my temerity in thinking I was ready to take charge of her so soon, and she caught on fire." When the commotion was over, I discovered that an electrical power box with enough current to light a city had short-circuited.

On to the next B-47 I went. "This 'girl' was more agreeable. We even managed to get the engines started. Then one of the 2,536,721 switches decided not to work. 'Quick!' someone shouted. 'Get the crew chief aboard before the idling engines burn up all our fuel!' The crew chief came aboard, Miss B-47 became cooperative, and once again all systems were go."

The takeoff was a thrill. Faster, faster, faster, faster . . . but we were still on the runway. I looked at the airspeed indicator. We were going 200 miles per hour, more than the top speed of most commercial airliners, and we hadn't left the ground! Then, just as the end of the runway appeared, we started to fly.

"Reluctant as our girl was to leave the ground, once airborne she climbed as if she were weightless. The altimeter raced around too fast to follow, and in a moment we were high above the clouds, going 600 miles per hour in a stillness like that of absolute space. I cannot describe it except with a worn-out cliché: 'It was like sitting on top of the world.' "

Then it was time to conduct the test. "Here was a little pipsqueak of an instrument, my Safe Flight Landing Speed Indicator, nestled amidst millions of dollars' worth of computers, radars,

gyroscopes, barometers, and gauges—electronics and superelec-
tronics. Would it work? Could little Safe Flight [my privately held
company] be right, and all our competitors [listed on the Stock Ex-
change], plus half of Wall Street, wrong? I set up the instrument,
told the pilot I was ready to land when he was, put on my crash helmet, and knocked on wood."

We came in for the landing. The speed was off the charts; the
plane was enormous, but the airport seemed to be shrinking. Just
when it was about to disappear entirely [we were going so fast, we
were passing right over it before we could land], the engine power
went on again; our pilot had decided not to land, for fear of not
being able to stop in time. We tried it again, and this time we made
it, but without much runway to spare. When we got out of the
plane, the pilot admitted he hadn't used my Indicator for the first
attempted landing, because he didn't trust it. When he did use it on
the second, it got us in safely!

"I was ready for a quick trip home in my nice, safe 200-mph
Beech, followed by dinner, bath, and bed. But Miss B-47 wasn't
through with me yet. I learned that the first plane, the one that
caught on fire, was almost fixed, and I was scheduled to align the
Indicator in it at 9:00 A.M. the next morning."

Needless to say, I didn't sleep too well, and I faced the flight
with some trepidation. By now familiar with the bailout procedure,
I quickly strapped on my assorted emergency gear, and, while the
jets were being fired up, got into the bomber's seat. Wedged between the quarter-million-dollar radar bombing computer (which
measures the Earth's rotation, etc.) and the side of the fuselage was
a comic book, left by my predecessor. Apparently, *he* wasn't wor-
ried about life insurance before each trip.

(continued)

Just as we were set to take off, a strong smell of burning rubber hit my nose. As smoke began pouring out of a hole in the complex bombing gear, I shouted into the radio, "Fire!" then scrambled out. Once more, we abandoned the ship to the fire crew; now more than before, I was ready for home, bath, and bed. But it still was not to be. The fire crew put out the flames, fixed the faulty circuit, and again we were "ready" to fly. This time, she got off.

Now my tune had changed: "Heaven again, or the nearest thing to it. Sitting in a plastic bubble, noiselessly speeding along at 600 mph, high above some messy snowstorms, I remembered a cold, gray, dismal Earth down below. Miss B-47 was moving fast enough to make it possible to eat breakfast in New York and Cairo on the same day."

When it was time to rejoin gray, dismal Earth, I followed the same procedure as the day before. Again the airport didn't seem large enough, and we had to make a go-around.

The B-47 has six engines and obviously it's preferable to have them all working when climbing away from a missed landing. But as the pilot applied the power, we felt a lurch, and the plane started to veer left. Two of the left engines had quit. The treetops were just below us, and the airplane was on the brink of turning over. The remaining four engines groaned and groaned and groaned. The pilot sweated, the co-pilot sweated, and I sweated.

But we survived, and I flew away in my nice, safe, comfortable 200-mph Beech writing this story and thinking to myself: home, dinner, bath, and bed.

SPREADING SIGHT

One fringe benefit of being an innovator is that you catch the attention of some interesting people, such as Walter Cronkite. Walter and I have shared many a delightful summer day on Martha's Vineyard. Like my wife, Walter is color-blind. When I showed him my wife's special glasses, he wanted a pair. I loaned him my wife's glasses, which he had duplicated. Later, on one of his broadcasts, Cronkite was able to report on the colors of the fish he had observed under water.

Another of my friends was the radio and television entertainer and acclaimed amateur pilot, Arthur Godfrey. A very active civilian pilot, he once told me that the Stall Warning Indicator I invented had saved his life. Through Arthur, I met renowned Air Force General Curtis E. LeMay. Arthur arranged for me to demonstrate my latest development in aviation technology—automatic throttles—to the general in Omaha, Nebraska.

Once in the air, I explained the equipment to General LeMay, then let him take the plane down. Back on the ground, he enthusiastically told everyone that with the aid of my equipment he had just landed a plane totally blind! My reward for that little demonstration was a 14-month military test of my invention, which proved to be very productive.

Thinking on Your Feet

In addition to being portable, inventorship is versatile—and makes you versatile, if you let it. Reminiscent of the anecdote Walter Cronkite told in the Foreword to this book (how his colleague used a Texaco credit card to gain entry to the Russian sector at the close of World War II), this story reiterates the importance of having inventorship capabilities at your disposal at all times.

About 25 years ago, my wife and I decided that it might be fun to circumnavigate South America in our airplane. We flew first to Bermuda, then to Grenada. Our next stop was to be Belém, Brazil. As they say, timing is everything, and sometimes it's bad timing. Already in the air en route to Brazil, we learned that our trip was coinciding with a revolutionary uprising, and the Belém airport had come under military control. It was too late to turn back—our fuel was too low by then for the return trip—so I made several attempts to gain permission to land, but received no response. Landing without permission, I feared, might put us in serious jeopardy.

Inventorship got us to the ground safely. I used my short-wave radio to call Kennedy Airport in New York and asked them to patch me through, by phone, to the airport in Belém. We received our clearance. (Calling Brazil from South America by way of New York may not seem so remarkable today when you can communicate with people all over the world via the Internet, but it was quite a novel thing to do a quarter-century ago.)

When we climbed out of the plane in Belém, we were greeted by a squad armed with machine guns, and were promptly placed under house arrest. We saw a lot of those guns over the next four hours, while we planned our next move. Neither of us knew Portuguese, but I was able to gesture well enough to convince our captors to take us back to the plane for an inspection that would show we were not a military threat. They searched our craft, and though they found nothing irregular, they still weren't satisfied. Somehow it became clear that if we could produce an insurance certificate, that would prove to them that our plane was not connected to the military in any way, and they would let us go.

We don't carry an insurance certificate in the plane, but, to buy time, my wife made a pretense of looking for it. In the process, she came across the plane's registration certificate. Pulling it out, she showed it to the chief and said, "Here it is." The certificate was in English, and so was unintelligible to our captors; in addition, it had that "official document look." It passed inspection as an insurance certificate, and the chief ordered our release. (See Figure 10.3.)

Thinking on your feet as a form of inventorship can get you out of many dilemmas. If you can't take the direct approach, try a workaround; if you're in a tight spot, find the crack and slip through it.

Figure 10.3 Sometimes, a good rule of inventorship is "Whatever works," as our phony insurance certificate proved.

EXERCISING INVENTORSHIP BY CHALLENGING YOURSELF

One of the best ways to exercise your inventorship capabilities is to look for ways to challenge yourself—just for the pure pleasure of making the effort. Find something you've always wanted to attempt, then go for it. You'll be pleasantly surprised at what you learn. I recall one of my favorite adventures while doing this.

The Fédération Aéronautique Internationale certifies world records in the field of aviation. I decided I wanted to set a distance record in the Beechcraft King Air my company was then using. The published range for it was 1,600 miles. I was determined to take it from Seattle to New York—almost twice that distance—without stopping for fuel.

According to Fédération rules, flyers may not modify the engine or add an extra fuel tank. In the model of plane I was flying, this meant that if I took off from Seattle, I could expect to run out of fuel over Minneapolis—not a welcoming prospect considering I had planned the trip for February.

I couldn't make the fuel tanks any bigger or more numerous, so what were my options? Just one, as far as I could see: somehow get more fuel into the tanks I had. With inventorship as my copilot, that turned out to be possible. A month before the flight, I put barrels of fuel in a cold-storage locker. Why? Because the low temperatures would shrink the volume of the fuel.

Also, when fuel is pumped into the tank of a plane, there's always an air pocket that is actually higher than the fuel cap, where, normally only air resides. I decided to make better use of that pocket: I filled my tanks from the drains in the bottom to make the fuel go all the way to the top. In short, when I filled 'er up, I really filled 'er up, to give me additional miles in the air.

I needed more, however, than just additional fuel. At 33,000 feet, I and my human copilot—my wife—would also need oxygen. But oxygen bottles add weight. So first I took seats out of my craft to make the plane lighter; then I added adjustable metering valves to the plane's oxygen line to make the vital gas last longer.

EXERCISING INVENTORSHIP BY CHALLENGING YOURSELF (CONTINUED)

Where there's a will, there's a way, as I learned when I became determined to set a distance record.

Next on my list was to modify the crew—my wife and myself. To preclude the need to run the cabin's heater, which uses up fuel, my wife and I wore warm, arctic clothing, so that we could make do with as little heat as possible.

I took one additional precautionary step. If you've flown commercially of late, you know that the key word in the phrase "estimated time of arrival" is "estimated." If we had to circle the airport before being cleared to land, we would waste a lot of our fuel. So I talked to the controllers a year before our flight, and, knowing that we were attempting to break a record, they agreed to give our flight priority, to avoid any delay in our landing.

(continued)

One aspect of our flight I couldn't control: the air itself. But I could control my response to it. When flying, the air can be your friend or your nemesis. Obviously, we wanted the air on our side when we were trying to take a plane twice as far on a tank of fuel as it's intended to go. I chose to make our flight in February because I knew the jet streams would give us the greatest tailwind. I did something else, too. The jet streams will often make a plane drift to the side, pushing it off course. To stay on course, a pilot normally "fights" the wind. But because my aim was to go a maximum distance on the least amount of fuel, I knew it would be better to go with the flow—literally. So, instead of correcting for jet-stream drift, we welcomed it. It pushed us off course and added a few fuel-consuming miles, but the boost we got from its tailwind saved us more fuel than the extra miles cost us. A straight line may be the shortest distance between two points, but that doesn't mean it's always the most efficient.

With all these measures and a few more, we took our little plane past what otherwise would have been a compulsory landing on the snowy plains of Minnesota. We made it safely to New York—with fuel to spare—and set the world's record for our class of aircraft. More importantly, though, we had a lot of fun in the process.

Having Fun with Inventorship

As I emphasized in Chapter 2, "The Age of Innovation," it's very important to encourage the inventive nature in children, and to support their imaginative play. As adults, it's equally important to indulge in the joys of playful inventorship, for it teaches some valuable lessons:

- We learn to laugh at ourselves.
- We learn that failure is survivable and can be a way station on the road to success.
- We learn that having fun with inventorship might not earn us a patent or a million dollars, but it will gift us with unforgettable memories.

The stories I share with you in this section emphasize the fun of leading an inventive life, and how ongoing those pleasures can be.

IMPRESSING DATES

When I was a student at Rutgers, the university was not yet coeducational, and I was dating a girl from the New Jersey College for Women, a couple of miles away. In those days, as you may have gathered from my horse-racing slot machine story in Chapter 6, I practiced my inventorship capabilities a great deal in an effort to impress coeds.

On this particular occasion, I had a date with my girlfriend to attend a fraternity dance. In those days, I was driving a Model A Ford, which, according to Mr. Ford's policy, was standard-issue black. But I was not quite as conservative as Henry. I decided to try and impress my date by giving our "carriage" for the evening a fancy paint job. The Rutgers school colors were scarlet and black, so that combination seemed a logical choice. The night before the dance, I gleefully set about painting the Ford's fenders scarlet, using only the light of a street lamp to guide my work. After I was finished, even in the dim lamplight, I thought the paint job looked terrific. Well satisfied with my idea and my work, I went home to bed.

My morning inspection of my handiwork made me anything but gleeful, however. My carefully painted scarlet fenders were covered with moths! Lured by the same light I had used to paint by, moths by the hundreds had landed on the sticky, wet paint, and become as trapped there as flies on flypaper. Now, instead of school-color fenders, the car sported moth fur. And I did not have enough time before the dance to undertake the arduous task of removing the moths from their scarlet resting place. (See Figure 10.4.)

As I saw it, my only option was to put on my best Joe College cool act, suit up in my scarlet jacket (at least I had

Figure 10.4 Practicing inventorship means that you must learn to bear your failures with a smile.

something that was scarlet) over white trousers, with white buck shoes, and go face my date in the Moth-Mobile. No question, she was shocked when she saw it, but I was on time, looked presentable, and had a corsage for her. With a confident smile, I got behind the wheel and we headed for the dance. So far, so good.

Unfortunately, my luck didn't hold, and the moth-covered fenders turned out to be the least of my concerns. The Ford's engine gave out halfway to our destination. Luckily, I knew what was wrong and how to fix it. The car's previous owner had apparently dropped a piece of rubber gasket into the gas tank, and it would occasionally sink into a position where it cut off the gravity flow of fuel to the engine.

To reach the gas cap, which was located right on the centerline of the hood, about six inches in front of the windshield, I had to take off my jacket and tie, stand on the front bumper, climb over the radiator, and spread-eagle myself on top of the hood, all the while taking care to avoid the insect-laden fenders. I put my mouth down on the cap, covering the vent opening. Then I gave the gas tank what amounted to a sort of CPR, sucking on the air vent, which brought air in through the bottom feed line and dislodged the piece of gasket.

Looking up and taking breath after doing this several times, I saw my date through the windshield only six inches from my face. (See Figure 10.5.) She was convulsed with laughter, tears were streaming down her face; and she could hardly breathe. She had been able to control herself when confronted by the Moth-Mobile, but the sight of her date spread across the hood of the car with his mouth on the gas cap was too much. She was in hysterics all the rest of the way to the dance; her tears streaked her mascara. And throughout the evening, she broke up anew every time she looked at me, which eventually broke up my confidence. It was, not surprisingly, my last date with this girl.

Figure 10.5   Sometimes a really inventive repair job causes those watching it to break, too—into uncontrollable laughter.

Undaunted by the Moth-Mobile fiasco, I continued to try and impress my dates with inventive ways to enhance our time together. On another occasion, a canoe was my vehicle of choice. When I was in school, Mill Pond, in New Brunswick, New Jersey, was a popular spot for college couples to go canoeing (read as a synonym for necking). It seemed to me that with a little inventorship I could make our evening something really special: I intended to set the mood with romantic music.

But this was long before the days of portable radios, so I had to make my own. I had a car radio in my Model A, powered by the car's battery. I removed both the radio and the battery, then reconnected them in the middle of our canoe. Result: Music over the water. Well, not quite. I had left the car's antenna behind, so it was more like static over the water.

But a practitioner of inventorship is never without ingenuity. I found that if I put a finger of one hand on the radio's antenna connection and a finger of the other hand in the water, I could substitute for the antenna, and have clear reception. (See Figure 10.6.) No doubt, the drawback to this solution is clear:

Figure 10.6 When practicing inventorship, on occasion you'll be faced with tough choices.

RECYCLING IS INVENTORSHIP

One summer, while vacationing on Chappaquiddick Island, near Martha's Vineyard, I learned that a new boat was going to be put into service for the ferry. Its predecessor was destined for the scrap heap. That seemed a terrible waste, so I volunteered to buy it for a dollar. Thereafter, its finest hours were as a party venue.

It was our tradition to have a big Fourth of July party at Edgartown, on the Vineyard, every year, and one summer we used the ferry as our dance floor. People were invited to wear costumes typically worn during the Roaring '20s—blazers and boaters, old-fashioned bathing dresses, and the like—and we decorated in the same style. The highlight was a huge weather balloon that we anchored

(continued)

RECYCLING IS INVENTORSHIP (CONTINUED)

to a basket on the deck blow, to resemble a hot-air balloon. In the basket, as if suspended from the balloon, was a three-man honky-tonk group playing our dance music.

At the end of the festivities, the trio played "Goodnight, Ladies" and when all the guests had left the ferry and were on the shore, we unslipped the moorings and floated away, just the two of us, continuing to dance to the strains of that song. Talk about an exit! It was a wonderful end to a wonderful party. The event was like something from an F. Scott Fitzgerald novel chronicling the extravagances of millionaires—but don't forget, I only paid a dollar for the ferry.

Inventorship includes the creative adaptation of items that might otherwise go to waste.

with both hands serving as a makeshift antenna, I had no hands to . . . well, you get the picture.

When I was 20, just two years out of college, and working, I had a date with a girl at her school, which was some distance away. I flew to her campus and landed at a field nearby.

(You may wonder how a 20-year old happened to have his own plane. By this time, I had already come up with my idea for the Stall Warning Indicator, and wanted to go into production with it, but I had no start-up capital. As a pilot, I reasoned that if I owned a plane, I could make money by flying short-hop charters, and I approached my mother with this idea. My mother, who had total faith in me, gave me the sum of $4,700, with which I was able to buy, from Tom Watson, later the president and CEO of IBM, his Ranger Fairchild 24, a fabric airplane—and a wonderful aircraft. I wish I had it today. With this airplane, I "hopped" enough passengers over a period of a few years to save $14,000, and with that money I started the Stall Warning system factory that became the Safe Flight Instrument Corporation, the company I still own today.)

My date and I enjoyed ourselves, and when it was time to say goodnight and wing my way home, I realized that I had a plane, but no airport, and therefore no lights to guide my takeoff. Worse, it was a moonless night, and I could not even see the field. I had noticed, however, when I'd landed earlier (while it was still light), that there were no obstructions on the field. I also knew it was long enough for my takeoff, if I could just take advantage of the field's full length.

My problem was how to maintain a straight path from one end of the field to the other, for if I veered off to one side, that would effectively shorten the length available to me. Luckily, I kept a flashlight in the cockpit. I gave it to my date and asked her to go to the far end of the field, turn on the flashlight, and point it at the plane. As she did that, I rolled the craft toward

Figure 10.7 Staying on the straight and narrow can require inventive thinking.

the flashlight. (See Figure 10.7.) I focused hard on that little pinpoint of light in front of me, and when the airspeed indicator showed that I had sufficient speed, I pulled back on the stick, and was soon safely on my way home.

Passing Inventorship On

If, as a young adult, you learn to have fun with inventorship, you will more easily be able to pass on to your own children that wonderful approach to life. Here's a sampling of my own efforts in this regard.

COMMUNICATING INVENTIVELY

When my children were young, they usually spent a part of each summer away at camp. Kids at camp are encouraged—really, required—to write home once a week, usually on Sunday, and that becomes a real chore, particularly after they've settled in, made friends, and are no longer homesick. So, to put a little fun back into the process, I decided to become an inventive communicator.

I sent my kids letters written backward. Unlike the greatest mirror-image writer, Leonardo da Vinci, I used a piece of carbon paper with the wrong side up. I also wrote letters using invisible ink, and let my kids figure out how to make the ink readable. Without a hint from me, they found that if they heated the paper over a candle, they could read my missives. On occasion, I even enclosed small treats—for example, a stick of chewing gum.

Another alternative I employed was to begin a letter in the middle of the paper, and write in an expanding circle until I reached the edges of the paper. This produced a spiral of writing that looked rather like a maze. (See Figure 10.8.)

The point is, by taking a chore and turning it into a game, I was able to communicate more effectively with my children. We replaced "duty" notes filled with such trite comments as, "I'm fine. Camp is fine. The food is lousy. Hope you're fine," with unusual, sometimes kooky letters. We all won!

TRAVELING INVENTIVELY

When our brood numbered three young children, I thought it would be a great idea to arrange a multifamily vacation to the

Figure 10.8 Taking the drudgery out of a chore is a form of inventorship, as my children found with inventive letter writing.

Bahamas. Now, I know what you're thinking, but I was determined to plan a trip guaranteed to prevent the typical hazards associated with traveling with young children: their boredom, followed quickly by restlessness, followed even more quickly by complaining and whining, concluding with equal amounts of boredom, weariness, and frustration on the part of the adults.

As part of my business, I owned an old DC-3 plane, and I decided to convert it temporarily into a flying motor home. I took the stops off the backs of all the side-by-side seats so that they could be laid back flat, then threw a lot of blankets and pillows on them, converting the area into one big communal bed. I added hammocks, slung from the hat racks. Now I had a plane that would sleep 24—for instance, five families, with 10 adults and 14 children. (See Figure 10.9.)

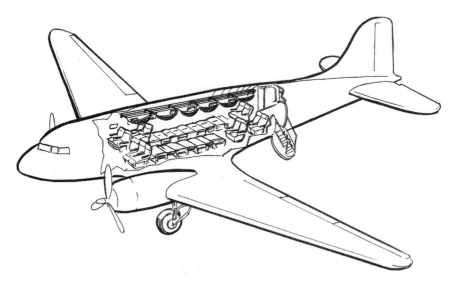

Figure 10.9   You can put the joy back into "joy ride" by using inventorship.

I told everyone to be onboard at 9:00 P.M. on departure "day," then just sat on the runway, while the kids were permitted to run around, excited at being up so late and on an airplane. After about an hour, they had let off enough steam so that by 10:00 P.M., they were sound asleep in the hammocks. The adults had settled down, too, and were either reading in the seats I had left upright, or starting to doze in the big communal bed. Everything was very calm and quiet as I took off, and then flew as slowly as I could, to make the trip last through the night. We arrived in the Bahamas right on schedule, just as the sun was coming up over the islands. It was a dramatic sight that none of us would forget.

I know that few families have a DC-3 at their disposal, and I admit that having one does make it easier to be innovative in regard to travel arrangements, but I also believe that anyone and everyone can make better use of what they do have to create comfort, joy, and fun in their lives. Just turn on inventorship; the trip will be more enjoyable, I promise.

TURNING THE WORLD UPSIDE DOWN AND INSIDE OUT

I mentioned in Chapter 3, "The Rule of Inventorship," the benefits of taking another—sometimes an opposing—view of a situation or a problem. I am such a proponent of this philosophy that I like to pass on the concept whenever I get the chance.

To that end, I used to display the globe in my home upside down. When people would ask why, I'd respond: "Upside down? Who says it's upside down?" Just because we're accustomed to seeing something a certain way doesn't necessarily make it the only way or the correct way. It all depends on how you look at it. I submit that turning the globe upside down is another way to discover the Earth.

Furthermore, as you have no doubt noticed, all globes show the Earth from the outside only. Did you ever wonder what our planet

TURNING THE WORLD UPSIDE DOWN AND INSIDE OUT (*CONTINUED*)

Inventorship challenges the conventional view. Why should north always be "up" on a globe?

looks like from the inside? What if we could get on the inside, and look out? I thought that would be an interesting point of view, so I papered the walls of a room with images of a transparent earth, as it might be seen from its center. Besides being a more interesting view, it's also a more accurate one, because it's free from the distortion you get with any top-down view of the continents.

Next time you're in Boston, go to the Christian Science Museum, where they have a wonderful globe on exhibit, and you can see exactly what I mean.

Assembling Pipe Dreams

In Chapter 2, "The Age of Innovation," I touched on a topic that bears emphasis: Parents do not have to spend money on the latest greatest toys for their children. In fact, to ensure that young imaginations are being stimulated, I'm a big believer in creating something from nothing or from the least likely items.

Yes, when you do this, some assembly *is* required, but it beats trying to figure out the confusing and often badly written instructions that come with most off-the-shelf, assembly-required toys. And when it's your own creation, the effort is a lot more satisfying and rewarding.

One of my favorite constructions was a jungle gym I formed from what I dubbed "Star Bars." I used 20-foot lengths of plastic PVC pipe, connected with rope through drilled holes, to form triangular shapes, in a design that looked straight from the brain of Buckminster Fuller. (See Figure 10.10.) My grandchildren were the first, and only, kids on their block to have a geodesic-domelike jungle gym.

The Star Bars were a big hit and have become the favorite neighborhood play center at several of my grandchildren's homes. They've also proven to be very safe, with a lot of advantages. They don't rust, for instance, and there are no splinters. They're easy to set up, move, or take down; and, because they're made of plastic, which has "give," they have yet to cause a single serious injury. Moreover, because the structure is made up of triangles, it's incredibly stable. There's no chance that it will tip over, no matter how many kids climb on, or what they're doing. At the same time, the plastic makes the whole thing so light that one person can lift it up and turn it upside down, so it can be used in another configuration. And, while the structure is fairly tall overall, it doesn't offer any height to fall from because the bars are criss-crossed all over in every direction. They provide

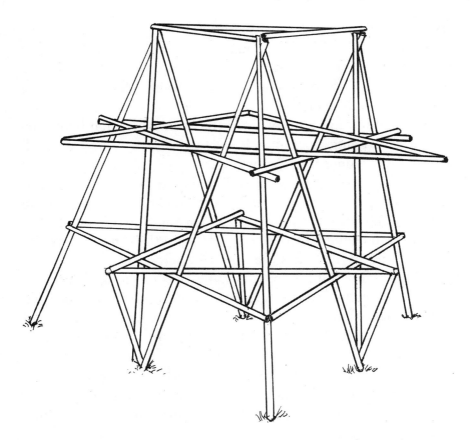

Figure 10.10 When you use your imagination, you can stimulate your children's minds at the same time.

almost infinite climbing routes, but no isolated high spots from which a child may take a dangerous fall.

Making the Inventorship Choice

My grandson Eric recently asked his father, "Is Poppy my grandfather?" His father replied, "Yes, he is; why do you ask?"

"Because that means that my grandfather is an inventor," Eric said happily, "and inventors can teach their grandchildren to do great things." Perhaps my grandson's simply stated conclusion explains better than I could why I believe so fervently in the power of inventorship.

I hope I haven't left you with the impression that inventorship means being unconventional for its own sake; as we all know, doing things in an unorthodox way is not always a good idea. My objective has been to demonstrate how inventorship encourages a free-ranging mind, one that can consider both the conventional and the unconventional, and then choose between them or merge them to create something new. Only the inventive mind really has that choice.

People who think conventionally become locked into orthodoxy; so-called nonconformists can become equally captive to their unconventionality. In contrast, inventorship offers all of us the choice of another way to solve our problems and to live our lives.

I think you can tell I enjoy my life, and wouldn't trade it for any other. It has been made richer by my practice of inventorship, and so, of course, it's a lifestyle I enthusiastically recommend for everyone.

I once asked a widow I knew: What would she look for in a man if ever she were to consider remarrying? "I would want our lives," she unhesitatingly replied, "never to be dull." Years later, long after I convinced that widow to become my bride, we both recalled with amusement her condition for remarriage. "Oh, what I wouldn't give now," she teased, "for a little dullness!"

I hope the insights I've provided in this book will lead you toward a life of inventorship. I can assure you, it will never be dull.

INDEX